SpringerBriefs in Mathematics

SpringerBriefs in Mathematics showcases expositions in all areas of mathematics and applied mathematics. Manuscripts presenting new results or a single new result in a classical field, new field, or an emerging topic, applications, or bridges between new results and already published works, are encouraged. The series is intended for mathematicians and applied mathematicians.

For further volumes:
http://www.springer.com/series/10030

Feng-Yu Wang

Harnack Inequalities for Stochastic Partial Differential Equations

Springer

Feng-Yu Wang
School of Mathematical Sciences
Beijing Normal University
Beijing, China, People's Republic

Department of Mathematics
Swansea University
Swansea, United Kingdom

ISSN 2191-8198 ISSN 2191-8201 (electronic)
ISBN 978-1-4614-7933-8 ISBN 978-1-4614-7934-5 (eBook)
DOI 10.1007/978-1-4614-7934-5
Springer New York Heidelberg Dordrecht London

Library of Congress Control Number: 2013944370

Mathematics Subject Classification (2010): 60H10, 60H15, 60H20

Printed on acid-free paper

Springer is part of Springer Science+Business Media (www.springer.com)

To my parents,
Shoujin Wang and Guijia Sang,
for their 80th birthdays

Preface

The key point of Harnack's inequality is to compare values at two different points for positive solutions of a partial differential equation. This inequality was introduced by Harnack [21] in 1887 for harmonic functions on a Euclidean space, and was generalized by Serrin [46] in 1955 and Moser [34] in 1961 to solutions of elliptic or parabolic partial differential equations. Among many other applications, Harnack's inequality was used by Li and Yau [26] in 1986 to derive explicit heat kernel estimates, and by Hamilton [20] in 1993 to investigate the regularity of Ricci flows, which was then used in Perelman's proof of the Poincaré conjecture. All these Harnack inequalities are, however, dimension-dependent and thus invalid for equations on infinite-dimensional spaces.

In this book we aim to present a self-contained account of Harnack inequalities and applications for the semigroup of solutions to stochastic functional/partial differential equations. Since the associated Fokker–Planck equations are partial differential equations on infinite-dimensional spaces, the Harnack inequalities we are going to investigate are dimension-free. This is essentially different from the above-mentioned classical Harnack inequalities. Moreover, the main tool in our study is a new coupling method (i.e., coupling by change of measure) rather than the usual maximum principle in the literature of partial differential equations and geometric analysis.

The book consists of four chapters. In Chap. 1, we introduce a general theory concerning dimension-free Harnack inequalities, which includes the main idea of establishing Harnack inequalities and derivative formulas using coupling by change of measure, derivative formulas using the Malliavin calculus, links of Harnack inequalities to gradient estimates, and various applications of Harnack inequalities. In Chap. 2, we establish the Harnack inequality with power and the log-Harnack inequality for the semigroup associated to a class of nonlinear stochastic partial differential equations, which include stochastic generalized porous media/fast-diffusion equations as typical examples. The main tool is the coupling by change of measure introduced in Chap. 1. In Chap. 3, we investigate gradient estimates and Harnack inequalities for semilinear stochastic partial differential equations using coupling by change of measure, gradient estimates, and finite-dimensional approximations.

Chapter 4 is devoted to gradient estimates and Harnack inequalities for the segment solution of stochastic functional differential equations, using coupling by change of measure and the Malliavin calculus. To save space, applications of Harnack and shift Harnack inequalities presented in Chap. 1 are not restated in the other three chapters for specific models.

In this book we consider only stochastic functional/partial differential equations driven by Brownian motions. But the general theory introduced in Chap. 1 works also for stochastic differential equations driven by Lévy noises or the fractional Brownian motions; see [16, 17, 62, 63, 67, 76] and references therein. Materials of the book are mainly organized from the author's recent publications, including joint papers with colleagues who are gratefully acknowledged for fruitful collaborations. In particular, I would like to mention the joint work [3] with Marc Arnaudon and Anton Thalmaier, where the coupling by change of measure was used for the first time to establish the dimension-free Harnack inequality.

I would like to thank Xiliang Fan and Shaoqin Zhang for reading earlier drafts of the book and making corrections. I would also like to thank my colleagues from the probability groups of Beijing Normal University and Swansea University, in particular Mu-Fa Chen, Wenming Hong, Niels Jacob, Zenghu Li, Eugene Lytvynov, Yonghua Mao, Aubrey Truman, Jiang-Lun Wu, Chenggui Yuan, and Yuhui Zhang. Their kind help and constant encouragement provided me with an excellent working environment.

Finally, financial support from the National Natural Science Foundation of China, Specialized Research Foundation for Doctoral Programs, the Fundamental Research Funds for the Central Universities, and the Laboratory of Mathematics and Complex Systems are gratefully acknowledged.

Beijing, China Feng-Yu Wang

Contents

1 A General Theory of Dimension-Free Harnack Inequalities 1
1.1 Coupling by Change of Measure and Applications 1
1.1.1 Harnack Inequalities and Bismut Derivative Formulas 2
1.1.2 Shift Harnack Inequalities and Integration by Parts Formulas 6
1.2 Derivative Formulas Using the Malliavin Calculus 8
1.2.1 Bismut Formulas 9
1.2.2 Integration by Parts Formulas 11
1.3 Harnack Inequalities and Gradient Inequalities 12
1.3.1 Gradient–Entropy and Harnack Inequalities 12
1.3.2 From Gradient–Gradient to Harnack Inequalities 16
1.3.3 L^2 Gradient and Harnack Inequalities 17
1.4 Applications of Harnack and Shift Harnack Inequalities 20
1.4.1 Applications of the Harnack Inequality 20
1.4.2 Applications of the Shift Harnack Inequality 25

2 Nonlinear Monotone Stochastic Partial Differential Equations 27
2.1 Solutions of Monotone Stochastic Equations 27
2.2 Harnack Inequalities for $\alpha \geq 1$ 30
2.3 Harnack Inequalities for $\alpha \in (0,1)$ 37
2.4 Applications to Specific Models 46
2.4.1 Stochastic Generalized Porous Media Equations 46
2.4.2 Stochastic p-Laplacian Equations 47
2.4.3 Stochastic Generalized Fast-Diffusion Equations 48

3 Semilinear Stochastic Partial Differential Equations 51
3.1 Mild Solutions and Finite-Dimensional Approximations 51
3.2 Additive Noise 57
3.2.1 Harnack Inequalities and Bismut Formula 57
3.2.2 Shift Harnack Inequalities and Integration by Parts Formula 62

3.3 Multiplicative Noise: The Log-Harnack Inequality ... 64
3.3.1 The Main Result ... 64
3.3.2 Application to White-Noise-Driven SPDEs ... 66
3.4 Multiplicative Noise: Harnack Inequality with Power ... 69
3.4.1 Construction of the Coupling ... 70
3.4.2 Proof of Theorem 3.4.1 ... 75
3.5 Multiplicative Noise: Bismut Formula ... 76

4 Stochastic Functional (Partial) Differential Equations ... 79
4.1 Solutions and Finite-Dimensional Approximations ... 79
4.1.1 Stochastic Functional Differential Equations ... 79
4.1.2 Semilinear Stochastic Functional Partial Differential Equations ... 83
4.2 Elliptic Stochastic Functional Partial Differential Equations with Additive Noise ... 85
4.2.1 Harnack Inequalities and Bismut Formula ... 85
4.2.2 Shift Harnack Inequalities and Integration by Parts Formulas ... 87
4.2.3 Extensions to Semilinear SDPDEs ... 90
4.3 Elliptic Stochastic Functional Partial Differential Equations with Multiplicative Noise ... 90
4.3.1 Log-Harnack Inequality ... 91
4.3.2 Harnack Inequality with Power ... 96
4.3.3 Bismut Formulas for Semilinear SDPDEs ... 103
4.4 Stochastic Functional Hamiltonian System ... 106
4.4.1 Main Result and Consequences ... 107
4.4.2 Proof of Theorem 4.4.1 ... 111
4.4.3 Proofs of Corollary 4.4.3 and Theorem 4.4.5 ... 115

Glossary ... 119

References ... 121

Index ... 125

Chapter 1
A General Theory of Dimension-Free Harnack Inequalities

1.1 Coupling by Change of Measure and Applications

The dimension-free Harnack inequality was first established in [50] for the heat semigroup on Riemannian manifolds with curvature bounded below. To derive the same type inequality on manifolds with unbounded below curvature, the coupling by change of measure was introduced in [3]. Then it was applied to the study of Harnack-type inequalities and derivative formulas for solutions of various stochastic equations; see, e.g., [4, 5, 10, 15, 19, 27, 31, 43, 53, 54, 57, 58, 61, 64, 65, 66, 68, 74]. In this section we explain the main idea for the study of Harnack inequalities and derivative formulas in an abstract framework.

Definition 1.1. Let μ and ν be two probability measures on a measurable space $(E,\mathscr{B})$, and let X,Y be two E-valued random variables on a complete probability space $(\Omega,\mathscr{F},\mathbb{P})$.

(i) If the distribution of X is μ, while under another probability measure $\mathbb{Q}$ on $(\Omega,\mathscr{F})$ the distribution of Y is ν, we call (X,Y) a *coupling by change of measure* for μ and ν with changed probability $\mathbb{Q}$.
(ii) If μ and ν are distributions of two stochastic processes with path space E, a coupling by change of measure (X,Y) for μ and ν is called a coupling by change of measure for these two processes. In this case, X and Y are called the marginal processes of the coupling (X,Y).

Let $\mathscr{B}(E)$, $\mathscr{B}_b(E)$, and $\mathscr{B}_b^+(E)$ denote the sets of all measurable, bounded measurable, and nonnegative bounded measurable functions on E. When E is a topological space, we take $\mathscr{B}$ to be the Borel σ-field, and denote by $C(E), C_b(E)$, and $C_b^+(E)$ the set of all continuous, bounded continuous, and nonnegative bounded continuous functions on E.

For a family of probability measures $\{\mu_x : x \in E\}$, we define

$$Pf(x) = \int_E f(y)\mu_x(\mathrm{d}y) =: \mu_x(f), \quad f \in \mathscr{B}_b(E), x \in E. \tag{1.1}$$

F.-Y. Wang, *Harnack Inequalities for Stochastic Partial Differential Equations*, SpringerBriefs in Mathematics, DOI 10.1007/978-1-4614-7934-5_1,

When Pf is measurable for $f \in \mathscr{B}_b(E)$, i.e., the family $\{\mu_x : x \in E\}$ is a transition probability measure, then P is a Markov operator on $\mathscr{B}_b(E)$, i.e., P is contractive and positivity-preserving, and $P1 = 1$.

When a family of stochastic processes $\{(X^x(t))_{t\geq 0} : x \in E\}$ measurable in x is involved, for instance $(X^x(t))_{t\geq 0}$ solves a stochastic differential equation with initial data x, let $\mu_x(t)$ be the distribution of $X^x(t)$. Then we define as in (1.1) a family of Markov operators $(P_t)_{t\geq 0}$. If the family of processes is Markovian, then $(P_t)_{t\geq 0}$ is a semigroup, i.e., $P_{t+s} = P_t P_s$ for $t,s \geq 0$.

In the remainder of the section, we will use coupling by change of measure to establish Harnack-type inequalities and derivative formulas of P.

1.1.1 Harnack Inequalities and Bismut Derivative Formulas

For a Markov operator P on $\mathscr{B}_b(E)$, the Harnack-type inequality considered in this book is of type

$$\Phi(Pf(x)) \leq (P\Phi(f)(y))\mathrm{e}^{\Psi(x,y)}, \quad x,y \in E, f \in \mathscr{B}_b^+(E), \tag{1.2}$$

where Φ is a nonnegative convex function on $[0,\infty)$ and Ψ is a nonnegative function on E^2. By Jensen's inequality, we may always take $\Psi(x,x) = 0$.

In this book we will mainly consider the following two typical choices of Φ:

(1) **(Harnack inequality with power)** Let $\Phi(r) = r^p$ for some $p > 1$. Then (1.2) reduces to

$$(Pf(x))^p \leq (Pf^p(y))\mathrm{e}^{\Psi(x,y)}, \quad x,y \in E, f \in \mathscr{B}_b^+(E). \tag{1.3}$$

This inequality, called the Harnack inequality with power p, was first found in [50] for diffusion semigroups with curvature bounded from below.

(2) **(Log-Harnack inequality)** Let $\Phi(r) = \mathrm{e}^r$. In this case we may use $\log f$ to replace f, so that (1.2) is equivalent to

$$P\log f(x) \leq \log Pf(y) + \Psi(x,y), \quad x,y \in E, f \in \mathscr{B}_b^+(E), f \geq 1. \tag{1.4}$$

Since the inequality does not change by multiplying f by a positive constant, it holds for all uniformly positive functions f. Moreover, using $f + \varepsilon$ to replace f and letting $\varepsilon \downarrow 0$, we may replace the condition $f \geq 1$ by $f \geq 0$. This inequality, called the log-Harnack inequality, was introduced in [56] for reflecting diffusion processes on manifolds with boundary and in [43] for semilinear SPDEs (stochastic partial differential equations) with multiplicative noise.

Theorem 1.1.1 (Harnack inequalities) *If there is a coupling by change of measure (X,Y) for μ_x and μ_y, with changed probability $\mathrm{d}\mathbb{Q} := R\mathrm{d}\mathbb{P}$, such that $X = Y$, $\mathbb{Q}$-a.s., then*

$$(Pf(y))^p \le (Pf^p)(x)\big(\mathbb{E}R^{\frac{p}{p-1}}\big)^{p-1}, \quad f \in \mathscr{B}_b^+(E),$$
$$P\log f(y) \le \log Pf(x) + \mathbb{E}(R\log R), \quad f \in \mathscr{B}_b^+(E), f > 0.$$

Proof. Since $Pf(x) = \mu_x(f) = \mathbb{E}f(X), Pf(y) = \mu_y(f) = \mathbb{E}(Rf(Y))$ and $X = Y, \mathbb{Q}$-a.s., by Hölder's inequality, we have

$$\begin{aligned}(Pf)^p(y) &= \big(\mathbb{E}(Rf(Y))\big)^p = \big(\mathbb{E}(Rf(X))\big)^p \\ &\le \big(\mathbb{E}f^p(X)\big)\big(\mathbb{E}R^{\frac{p}{p-1}}\big)^{p-1} = \big(Pf^p(x)\big)\big(\mathbb{E}R^{\frac{p}{p-1}}\big)^{p-1}.\end{aligned}$$

Thus, the first inequality holds. Next, by Young's inequality we have

$$P\log f(y) = \mathbb{E}(R\log f(X)) \le \log \mathbb{E}f(X) + \mathbb{E}(R\log R) = \log Pf(x) + \mathbb{E}(R\log R).$$

So the second inequality holds. □

Next, we show that the coupling by change of measure can also be used to establish a Bismut-type derivative formula of P.

Theorem 1.1.2 (Bismut formula) *Let $\gamma : [0, r_0] \to E$ with $r_0 > 0$ be a curve on E, and let X be an E-valued random variable with distribution $\mu_{\gamma(0)}$. If for every $\varepsilon \in (0, r_0)$ there exists a coupling by change of measure (X, X^ε) with changed probability $\mathrm{d}\mathbb{Q}_\varepsilon := R_\varepsilon \mathrm{d}\mathbb{P}$ for $\mu_{\gamma(0)}$ and $\mu_{\gamma(\varepsilon)}$ such that $X = X^\varepsilon, \mathbb{Q}_\varepsilon$-a.s. and*

$$M := \lim_{\varepsilon \to 0} \frac{R_\varepsilon - 1}{\varepsilon}$$

exists in $L^1(\mathbb{P})$, then

$$\frac{\mathrm{d}}{\mathrm{d}\varepsilon} Pf(\gamma(\varepsilon))\Big|_{\varepsilon=0} = \mathbb{E}\big[Mf(X)\big], \quad f \in \mathscr{B}_b(E).$$

Proof. Simply note that under the given conditions, we have

$$\begin{aligned}&\lim_{\varepsilon \to 0} \frac{Pf(\gamma(\varepsilon)) - Pf(\gamma(0))}{\varepsilon} = \lim_{\varepsilon \to 0} \frac{\mathbb{E}[R_\varepsilon f(X^\varepsilon)] - \mathbb{E}f(X)}{\varepsilon} \\ &= \lim_{\varepsilon \to 0} \frac{1}{\varepsilon} \mathbb{E}\big[f(X)(R_\varepsilon - 1)\big] = \mathbb{E}[Mf(X)].\end{aligned}$$

□

Below we present a simple example to illustrate the above two theorems. Consider the following SDE (stochastic differential equation) on $\mathbb{R}^d$:

$$\mathrm{d}X(t) = b(X(t))\mathrm{d}t + \mathrm{d}B(t), \tag{1.5}$$

where $B(t)$ is d-dimensional Brownian motion and $b : \mathbb{R}^d \to \mathbb{R}^d$ is continuous such that

$$\langle b(x) - b(y), x - y\rangle \le K|x - y|^2, \quad x, y \in \mathbb{R}^d \tag{1.6}$$

holds for some constant $K \in \mathbb{R}$. It is well known that for every $x \in \mathbb{R}^d$, (1.5) has a unique solution $\{X^x(t)\}_{t\geq 0}$ starting from x and that the solution is nonexplosive. Then the associated Markov semigroup $\{P_t\}_{t\geq 0}$ is given by

$$P_t f(x) = \mathbb{E} f(X^x(t)), \ \ x \in \mathbb{R}^d, f \in \mathscr{B}_b(\mathbb{R}^d), t \geq 0.$$

Let $t > 0$ and $x \in \mathbb{R}^d$ be fixed. To apply Theorem 1.1.1 to P_t, we let $X = X^x$ and let Y solve the equation

$$\mathrm{d}Y(s) = \left(b(Y(s)) + \eta(s) 1_{[0,\tau)}(s) \cdot \frac{X(s) - Y(s)}{|X(s) - Y(s)|} \right) \mathrm{d}s + \mathrm{d}B(s), \ \ Y(0) = y,$$

where

$$\tau := \inf\{s \geq 0 : X(s) = Y(s)\}$$

is the coupling time and $\eta \in C([0,\infty))$ is to be determined. It is easy to see that this equation has a unique solution up to the coupling time τ. Letting $Y(s) = X(s)$ for $s \geq \tau$, we obtain a solution $Y(s)$ for all $s \geq 0$. We will then choose η such that $\tau \leq t$, i.e., $X(t) = Y(t)$.

By (1.6), we obtain

$$\mathrm{d}|X(s) - Y(s)| \leq \{K|X(s) - Y(s)| - \eta(s)\} \mathrm{d}s, \ \ s < \tau.$$

Then

$$\mathrm{e}^{-K(\tau \wedge t)} |X(t \wedge \tau) - Y(t \wedge \tau)| \leq |x - y| - \int_0^{t \wedge \tau} \mathrm{e}^{-Ks} \eta(s) \mathrm{d}s.$$

Taking

$$\eta(s) = \frac{|x - y| \mathrm{e}^{-Ks}}{\int_0^t \mathrm{e}^{-2Ks} \mathrm{d}s}, \ \ s \in [0,t],$$

we see that $|x - y| - \int_0^t \mathrm{e}^{-Ks} \eta(s) \mathrm{d}s = 0$, so that $\tau \leq t$, and thus $X(t) = Y(t)$, as required. To see that (X,Y) is a coupling by change of measure for X^x and X^y, let

$$R = \exp\left[- \int_0^\tau \frac{\eta(s)}{|X(s) - Y(s)|} \langle X(s) - Y(s), \mathrm{d}B(s) \rangle - \frac{1}{2} \int_0^\tau |\eta(s)|^2 \mathrm{d}s \right].$$

By Girsanov's theorem, under the probability $\mathrm{d}\mathbb{Q} := R \mathrm{d}\mathbb{P}$, the process

$$\tilde{B}(s) := B(s) + \int_0^{s \wedge \tau} \eta(r) \cdot \frac{X(r) - Y(r)}{|X(r) - Y(r)|} \mathrm{d}r, \ \ s \geq 0$$

is a d-dimensional Brownian motion. Reformulating the equation for $Y(s)$ as

$$\mathrm{d}Y(s) = b(Y(s)) + \mathrm{d}\tilde{B}(s), \ \ Y(0) = y,$$

we see that the distribution of Y under $\mathbb{Q}$ coincides with that of X^y under $\mathbb{P}$. Therefore, $(X(t), Y(t))$ is a coupling by change of measure for $X^x(t)$ and $X^y(t)$ with changed probability $\mathbb{Q}$ such that $X(t) = Y(t)$. According to Theorem 1.1.1, for every $p > 1$ we have

$$P_t f(y) \le \big(P_t f^p(x)\big)^{\frac{1}{p}} \big(\mathbb{E} R^{\frac{p}{p-1}}\big)^{\frac{p-1}{p}}, \quad f \in \mathscr{B}_b^+(\mathbb{R}^d).$$

Since $\tau \le t$ and the definition of η imply

$$\mathbb{E} R^{\frac{p}{p-1}} \le \exp\left[\frac{p}{2(p-1)^2}\int_0^t |\eta(s)|^2 \mathrm{d}s\right] = \exp\left[\frac{pK|x-y|^2}{(p-1)^2(1-\mathrm{e}^{-2Kt})}\right],$$

we obtain the following Harnack inequality with power p:

$$(P_t f(x))^p \le (P_t f^p(y)) \exp\left[\frac{pK|x-y|^2}{(p-1)(1-\mathrm{e}^{-2Kt})}\right], \quad f \in \mathscr{B}_b^+(\mathbb{R}^d).$$

Noting that

$$\mathbb{E} R \log R = \mathbb{E}_{\mathbb{Q}} \log R = \frac{1}{2}\mathbb{E}_{\mathbb{Q}} \int_0^\tau |\eta(s)|^2 \mathrm{d}s \le \frac{K|x-y|^2}{1-\mathrm{e}^{-2Kt}},$$

we have the following log-Harnack inequality:

$$P_t \log f(y) \le \log P_t f(x) + \frac{K|x-y|^2}{1-\mathrm{e}^{-2Kt}}, \quad f \in \mathscr{B}_b^+(\mathbb{R}^d), f > 0.$$

Next, to derive the Bismut formula, we assume further that b is Lipschitz continuous. For $v \in \mathbb{R}^d$ and $\varepsilon > 0$, let Y^ε solve the equation

$$\mathrm{d}Y^\varepsilon(s) = \Big(b(X(s)) - \frac{\varepsilon v}{t}\Big)\mathrm{d}s + \mathrm{d}B(s), \quad Y^\varepsilon(0) = x + \varepsilon v.$$

Then $Y^\varepsilon(s) = X(s) + \frac{\varepsilon(t-s)}{t} v, \ s \in [0,t]$. Let

$$R_\varepsilon = \exp\left[\int_0^t \Big\langle \frac{\varepsilon v}{t} + b(Y^\varepsilon(s)) - b(X(s)), \mathrm{d}B(s)\Big\rangle \right.$$
$$\left. - \frac{1}{2}\int_0^t \Big|\frac{\varepsilon v}{t} - b(X(s)) + b(Y^\varepsilon(s))\Big|^2 \mathrm{d}s\right].$$

It is easy to see that under $\mathrm{d}\mathbb{Q}_\varepsilon := R_\varepsilon \mathrm{d}\mathbb{P}$, the distribution of Y^ε coincides with that of $X^{x+\varepsilon v}$ under $\mathbb{P}$. Therefore, by Theorem 1.1.2, we obtain

$$\begin{aligned}\nabla_v P_t f(x) &= \mathbb{E}\Big(f(X(t)) \lim_{\varepsilon \to 0} \frac{R_\varepsilon - 1}{\varepsilon}\Big)\\ &= \mathbb{E}\Big(\frac{f(X(t))}{t}\int_0^t \big\langle v + (t-s)(\nabla_v b)(X(s)), \mathrm{d}B(s)\big\rangle\Big), \quad f \in \mathscr{B}_b(\mathbb{R}^d);\end{aligned}$$

using Malliavin calculus. See also (1.13) below. We remark that this formula is a different version of the original Bismut formula derived using the Malliavin calculus in [9] and using a martingale argument in [14]:

$$\nabla_v P_t f(x) = \mathbb{E}\left(\frac{f(X(t))}{t}\int_0^t \left\langle \nabla_v (X^{\cdot}(s))(x), \mathrm{d}B(s)\right\rangle\right), \quad f \in \mathscr{B}_b(\mathbb{R}^d). \tag{1.7}$$

See (1.14) below for a more general version due to Thalmaier [49].

1.1.2 Shift Harnack Inequalities and Integration by Parts Formulas

Theorem 1.1.3 *Let E be a Banach space and let $x, v \in E$ be fixed.*

(1) *For every coupling by change of measure (X,Y) with changed probability $\mathbb{Q} = R\mathbb{P}$ for μ_x and μ_x such that $Y = X + v, \mathbb{Q}$-a.s., there hold the shift Harnack inequality*

$$|Pf(x)|^p \le P\{|f|^p(v+\cdot)\}(x)\big(\mathbb{E}R^{\frac{p}{p-1}}\big)^{p-1}, \quad f \in \mathscr{B}_b(E),$$

and the shift log-Harnack inequality

$$P\log f(x) \le \log P\{f(v+\cdot)\}(x) + \mathbb{E}(R\log R), \quad f \in \mathscr{B}_b(E), f > 0.$$

(2) *Let $(X, Y^\varepsilon), \varepsilon \in [0,1]$, be a family of couplings by change of measure for μ_x and μ_x with changed probability $\mathbb{Q}_\varepsilon = R_\varepsilon \mathbb{P}$ such that*

$$Y^\varepsilon = X + \varepsilon v, \ \mathbb{Q}_\varepsilon\text{−a.s.}, \varepsilon \in (0,1].$$

If $N := \lim_{\varepsilon\to 0} \frac{1-R_\varepsilon}{\varepsilon}$ exists in $L^1(\mathbb{P})$, then

$$P(\nabla_v f)(x) = \mathbb{E}\big\{f(X)N\big\}, \quad f, \nabla_v f \in \mathscr{B}_b(E).$$

Proof. The proof is similar to that introduced above for the Harnack inequality and Bismut formula.

(1) Note that $Pf(x) = \mathbb{E}\{Rf(Y)\} = \mathbb{E}\{Rf(X+v)\}$. We have

$$|Pf(x)|^p \le \big(\mathbb{E}|f|^p(X+v)\big)\big(\mathbb{E}R^{\frac{p}{p-1}}\big)^{p-1} = P\{|f|^p(v+\cdot)\}(x)\big(\mathbb{E}R^{\frac{p}{p-1}}\big)^{p-1}.$$

Similarly, for positive f,

$$\begin{aligned} P\log f(x) &= \mathbb{E}\{R\log f(X+v)\} \\ &\le \log \mathbb{E} f(X+v) + \mathbb{E}(R\log R) = \log P\{f(v+\cdot)\}(x) + \mathbb{E}(R\log R). \end{aligned}$$

(2) Noting that $Pf(x) = \mathbb{E}\big\{R_\varepsilon f(Y^\varepsilon)\big\} = \mathbb{E}\big\{R_\varepsilon f(X+\varepsilon v)\big\}$, we obtain

$$0 = \lim_{\varepsilon \to 0} \frac{1}{\varepsilon} \mathbb{E}\{R_\varepsilon f(X+\varepsilon v) - f(X)\} = P(\nabla_v f)(x) - \mathbb{E}\{f(X)N\},$$

provided $N := \lim_{\varepsilon \to 0} \frac{1-R_\varepsilon}{\varepsilon}$ exists in $L^1(\mathbb{P})$. □

Let us again consider (1.5) to illustrate Theorem 1.1.3. Let $X(s)$ solve (1.5) with $X(0) = x$. Let $Y(s) = X(s) + \frac{s}{t}v$. Then

$$\mathrm{d}Y(s) = b(Y(s))\mathrm{d}s + \mathrm{d}\tilde{B}(s), \quad Y(0) = X(0) = x,$$

where

$$\tilde{B}(s) := B(s) + \int_0^s \Big(\frac{v}{t} + b(X(r)) - b(Y(r))\Big)\mathrm{d}r$$

is a d-dimensional Brownian motion under the probability $\mathbb{Q} := R\mathbb{P}$ for

$$R := \exp\bigg[-\int_0^t \Big\langle \frac{v}{t} + b(X(r)) - b(Y(r)), \mathrm{d}B(r)\Big\rangle - \frac{1}{2}\int_0^t \Big|\frac{v}{t} + b(X(r)) - b(Y(r))\Big|^2 \mathrm{d}r\bigg].$$

That is, $(X(t), Y(t))$ is a coupling by change of measure for $X^x(t)$ and $X^x(t)$ with changed probability $\mathbb{Q}$ such that $Y(t) = X(t) + v$. Assuming $|b(x) - b(y)| \le c|x-y|$, we see that

$$\Big|\frac{v}{t} + b(X(r)) - b(Y(r))\Big| \le \frac{1+cr}{t}|v|.$$

Then for $p > 1$ and $f \in \mathscr{B}_b^+(\mathbb{R}^d)$,

$$\begin{aligned}(P_t f)^p(x) &\le (P_t f^p(v+\cdot))(x)\big(\mathbb{E}R^{\frac{p}{p-1}}\big)^{p-1}\\ &\le (P_t f^p(v+\cdot))(x) \exp\bigg[\frac{p|v|^2}{2(p-1)}\Big(\frac{1}{t} + c + \frac{c^2 t}{3}\Big)\bigg].\end{aligned}$$

To derive the integration by parts formula, let

$$Y^\varepsilon(s) = X(s) + \frac{\varepsilon s}{t}v, \quad s \ge 0.$$

Assume, for instance, that $b \in C_b^1(\mathbb{R}^d; \mathbb{R}^d)$. Then Theorem 1.1.3(2) applies to

$$\begin{aligned}R_\varepsilon = \exp\bigg[&-\int_0^t \Big\langle \frac{\varepsilon}{t}v + b(X(r)) - b(Y^\varepsilon(r)), \mathrm{d}B(r)\Big\rangle\\ &-\frac{1}{2}\int_0^t \Big|\frac{\varepsilon}{t}v + b(X(r)) - b(Y^\varepsilon(r))\Big|^2 \mathrm{d}r\bigg].\end{aligned}$$

It is easy to see that

$$\frac{\mathrm{d}}{\mathrm{d}\varepsilon}R_\varepsilon\Big|_{\varepsilon=0} = \frac{1}{t}\int_0^t \langle s(\nabla_v b)(X(s)) - v, \mathrm{d}B(s)\rangle.$$

Therefore,

$$\mathbb{E}(\nabla_v f)(X(t)) = \mathbb{E}\left[\frac{f(X(t))}{t}\int_0^t \langle v - s(\nabla_v b)(X(s)), \mathrm{d}B(s)\rangle\right]. \tag{1.8}$$

1.2 Derivative Formulas Using the Malliavin Calculus

Let $(\mathbb{H}, \langle\cdot,\cdot\rangle, |\cdot|)$ be a separable Hilbert space and let $W := \{W(t)\}_{t\in[0,T]}$ be a cylindrical Brownian motion on $\mathbb{H}$ with respect to a complete filtered probability space $(\Omega, \mathscr{F}, \{\mathscr{F}_t\}_{t\geq 0}, \mathbb{P})$; that is, for every orthonormal elements $\{e_1, \cdots, e_n\}$ in $\mathbb{H}$, $(\langle W, e_1\rangle, \cdots, \langle W, e_n\rangle)$ is a Brownian motion on $\mathbb{R}^n$. One may formally write

$$W(t) = \sum_{n=1}^{\infty} B_n(t) e_n, \quad t \in [0,T],$$

for an orthonormal basis $\{e_n\}_{n\geq 1}$ of $\mathbb{H}$ and a family of independent one-dimensional Brownian motions $\{B_n\}_{n\geq 1}$. Thus, if $\mathbb{H}$ is infinite-dimensional, W is not a process on $\mathbb{H}$. It is convenient to realize W as a continuous process on an enlarged Hilbert space, for instance on $\tilde{\mathbb{H}}$, the completion of $\mathbb{H}$ under the inner product

$$\langle x, y\rangle_{\tilde{\mathbb{H}}} := \sum_{i=1}^{\infty} 2^{-i}\langle x, e_i\rangle\langle y, e_i\rangle, \quad x, y \in \mathbb{H}.$$

Let μ_T be the distribution of W, which is then a probability measure on the path space

$$\tilde{\mathscr{C}} := \{\gamma \in C([0,T]; \tilde{\mathbb{H}}),\ \gamma(0) = 0\},$$

equipped with the Borel σ-field of the uniform norm.

To introduce the Malliavin derivative of a functional of $W_{[0,T]} := \{W(t)\}_{t\in[0,T]}$, let

$$\mathbb{H}^1 = \left\{h \in C([0,T]; \mathbb{H}) :\ h(0) = 0, \|h\|_{\mathbb{H}^1}^2 := \int_0^T |h'(t)|^2 \mathrm{d}t < \infty\right\}$$

be the tangent space over $\tilde{\mathscr{C}}$, which is known as Cameron–Martin space. It is again a Hilbert space with inner product $\langle h_1, h_2\rangle_{\mathbb{H}^1} := \int_0^T \langle h_1'(t), h_2'(t)\rangle \mathrm{d}t$. We call a function $F \in L^2(\tilde{\mathscr{C}}; \mu_T)$ differentiable if

$$D_h F := \lim_{\varepsilon\to 0} \frac{F(\cdot + \varepsilon h) - F}{\varepsilon}$$

exists in $L^2(\tilde{\mathscr{C}}; \mu_T)$ for every $h \in \mathbb{H}^1$. If, moreover, $h \mapsto D_h F$ is bounded from $\mathbb{H}^1$ to $L^2(\tilde{\mathscr{C}}; \mu_T)$, then by the Riesz representation theorem, there exists a unique element $DF \in L^2(\tilde{\mathscr{C}} \to \mathbb{H}^1; \mu_T)$ such that

$$\langle h, DF\rangle_{\mathbb{H}^1} = D_h F, \quad \mu_T\text{-a.s.}$$

In this case, we denote $F \in \mathscr{D}(D)$ and call DF the Malliavin gradient of F. Then $(D, \mathscr{D}(D))$ is a closed operator from $L^2(\tilde{\mathscr{C}}; \mu_T)$ to $L^2(\tilde{\mathscr{C}} \to \mathbb{H}^1; \mu_T)$, whose adjoint operator is denoted by $(D^*, \mathscr{D}(D^*))$. Thus, we have the following integration by parts formula:

$$\mathbb{E}(D_h F)(W) = \int_{\mathscr{C}} D_h F \mathrm{d}\mu_T = \int_{\mathscr{C}} F D^* h \mathrm{d}\mu_T = \mathbb{E}(F D^* h)(W) \tag{1.9}$$

for $F \in \mathscr{D}(D), h \in \mathscr{D}(D^*)$. If $h \in L^2(\tilde{\mathscr{C}} \to \mathbb{H}^1; \mu_T)$ such that $h(W)(\cdot)$ is adapted, i.e., $h(W)(t)$ is $\mathscr{F}_t$-measurable for $t \in [0,T]$, then $h \in \mathscr{D}(D^*)$ and (see [32, 35])

$$(D^* h)(W) = \int_0^T \langle \{h(W)\}'(t), \mathrm{d}W(t) \rangle. \tag{1.10}$$

Now let $\{X^x\}_{x \in \mathbb{H}}$ be a family of $\mathbb{H}$-valued measurable functionals of W such that

$$Pf(x) := \mathbb{E} f(X^x), \quad f \in \mathscr{B}_b(\mathbb{H}), \ x \in \mathbb{H},$$

gives rise to a Markov operator on $\mathscr{B}_b(\mathbb{H})$.

In the following two subsections, we will use the Malliavin calculus to study derivative formulas of P. These results will be then illustrated by the Markov semigroup associated to (1.5). Although for this simple model these two different arguments lead to the same derivative formula, we will see in the other chapters of the book that each of these arguments has its own advantages. In particular, when the Harnack inequality is involved, the coupling argument is more straightforward, and it allows stochastic equations having singular coefficients, but when the derivative formula is involved, the other argument is normally more convenient.

1.2.1 Bismut Formulas

The following result is a standard application of the Malliavin calculus.

Theorem 1.2.1 *Let $x, v \in \mathbb{H}$ be such that*

$$\nabla_v X^x := \lim_{\varepsilon \to 0} \frac{X^{x+\varepsilon v} - X^x}{\varepsilon}$$

exists in $L^1(\mathbb{P})$. If $X^x \in \mathscr{D}(D)$ and there exists $h \in \mathscr{D}(D^*)$ such that

$$\nabla_v X^x = D_h X^x, \tag{1.11}$$

then

$$\nabla_v Pf(x) = \mathbb{E}\big(f(X^x) D^* h\big), \quad f \in \mathscr{B}_b(\mathbb{H}).$$

Proof. By the monotone class theorem, it suffices to prove the result for f having uniformly bounded directional derivatives. In this case, it follows from the chain rule, (1.11), and (1.9) that

$$\nabla_v Pf(x) = \mathbb{E}\{(\nabla_{\nabla_v X^x} f)(X^x)\} = \mathbb{E}\{(\nabla_{D_h X^x} f)(X^x)\}$$
$$= \mathbb{E}\{D_h(f(X^x))\} = \mathbb{E}\{f(X^x)D^*h\}.$$

□

Now we apply Theorem 1.2.1 to $(P_t)_{t\geq 0}$ associated to (1.5) with b having bounded gradient. In this case, we set $\mathbb{H} = \mathbb{R}^d$ and $X^x = X^x(T)$ for a fixed time $T > 0$. By (1.5), we have

$$\begin{aligned} \mathrm{d}\nabla_v X^x(t) &= \big(\nabla_{\nabla_v X^x(t)} b\big)(X^x(t))\mathrm{d}t, \ \ \nabla_v X^x(0) = v, \\ \mathrm{d}D_h X^x(t) &= \big(\nabla_{D_h X^x(t)} b\big)(X^x(t))\mathrm{d}t + h'(t)\mathrm{d}t, \ \ D_h X^x(0) = 0. \end{aligned} \tag{1.12}$$

Then $v(t) := \nabla_v X^x(t) - D_h X^x(t)$ solves the equation

$$\mathrm{d}v(t) = \big(\nabla_{v(t)} b\big)(X^x(t))\mathrm{d}t - h'(t)\mathrm{d}t, \ \ v(0) = v.$$

Taking

$$h(t) = \frac{t}{T}v + \int_0^t \big(\nabla_{v(s)} b\big)(X^x(s))\mathrm{d}s,$$

we obtain

$$v(t) = \frac{T-t}{T}v, \ \ t \in [0,T].$$

In particular, $v(T) = 0$, i.e., $\nabla_v X^x(T) = D_h X^x(T)$ as required. So

$$h(t) := \frac{t}{T}v + \int_0^t \frac{T-s}{T}(\nabla_v b)(X^x(s))\mathrm{d}s, \ \ t \in [0,T],$$

is a good choice for applying Theorem 1.2.1. Therefore, by Theorem 1.2.1 and (1.10) we obtain

$$\nabla_v P_T f(x) = \mathbb{E}\bigg(\frac{f(X^x(T))}{T}\int_0^T \langle v + (T-t)(\nabla_v b)(X^x(t)), \mathrm{d}B(t)\rangle\bigg) \tag{1.13}$$

as in the last section using coupling by change of measure.

To derive the classical Bismut formula (1.7), let $\ell \in C^1([0,T])$ with $\ell(0) = 0$ and $\ell(T) = 1$. Take

$$h(t) = \int_0^t \ell'(s)\nabla_v X^x(s)\mathrm{d}s, \ \ t \in [0,T].$$

Then $h \in \mathscr{D}(D^*)$, and by (1.12), $v(t) := \ell(t)\nabla_v X^x(t)$ solves the equation

$$\mathrm{d}v(t) = (\nabla_{v(t)} b)(X^x(t))\mathrm{d}t + h'(t)\mathrm{d}t, \ \ t \in [0,T], v(0) = 0.$$

Since $D_h X^x(t)$ solves the same equation, by the uniqueness of the solution we have

$$D_h X^x(T) = v(T) = \nabla_v X^x(T).$$

Therefore, by Theorem 1.2.1 and (1.10) we obtain

$$\nabla_v P_T f(x) = \mathbb{E}\left(f(X^x(T)) \int_0^T \ell'(t) \langle \nabla_v X^x(t), \mathrm{d}B(t) \rangle \right). \tag{1.14}$$

This formula is due to [49], and it recovers (1.7) by taking $\ell(t) = \frac{t}{T}$.

1.2.2 Integration by Parts Formulas

Theorem 1.2.2 *Let $x, v \in \mathbb{H}$. If $X^x \in \mathscr{D}(D)$ and there exists $h \in \mathscr{D}(D^*)$ such that*

$$D_h X^x = v, \tag{1.15}$$

then

$$P(\nabla_v f)(x) = \mathbb{E}\big(f(X^x) D^* h\big), \quad f \in C_b^1(\mathbb{H}).$$

Proof. By the chain rule and (1.9), we have

$$\begin{aligned} P(\nabla_v f)(x) &= \mathbb{E}\{(\nabla_v f)(X^x)\} = \mathbb{E}\{(\nabla_{D_h X^x} f)(X^x)\} \\ &= \mathbb{E}\{D_h(f(X^x))\} = \mathbb{E}\{f(X^x) D^* h\}. \end{aligned}$$

□

To apply Theorem 1.2.2 to P_T associated to (1.5), we need to find $h \in \mathscr{D}(D^*)$ such that $D_h X^x(T) = v$. Recall that

$$\mathrm{d}D_h X^x(t) = \big(\nabla_{D_h X^x(t)} b\big)(X^x(t))\mathrm{d}t + h'(t)\mathrm{d}t, \quad D_h X^x(0) = 0.$$

So, letting

$$h(t) = \frac{t}{T} v - \int_0^t \big(\nabla_{D_h X^x(s)} b\big)(X^x(s))\mathrm{d}s,$$

we obtain

$$D_h X^x(t) = \frac{t}{T} v, \quad t \in [0, T].$$

In particular, $D_h X^x(T) = v$, as required, and

$$h(t) = \frac{t}{T} v - \int_0^t \frac{s}{T} (\nabla_v b)(X^x(s))\mathrm{d}s, \quad t \in [0, T].$$

So we obtain (1.8) from Theorem 1.2.2 and (1.10).

We would like to mention that the integration by parts formula for the heat semigroup on a compact Riemannian manifold was first established by Driver in [13], in which both Ricci curvature and its first-order derivatives are involved.

1.3 Harnack Inequalities and Gradient Inequalities

In this section we aim to characterize links between Harnack-type inequalities and gradient estimates for a Markov operator P on $\mathscr{B}_b(E)$, where (E,ρ) is a geodesic space, and to show that the Harnack inequality with power implies the log-Harnack inequality when (E,ρ) is a length space. Most results in this section are essentially due to [4,43,56,60,61].

Recall that a metric space (E,ρ) is called a *geodesic space* if for every $x,y \in E$, there exists a map $\gamma\colon [0,1] \to E$ such that $\gamma(0)=x, \gamma(1)=y$ and $\rho(\gamma(s),\gamma(t)) = |t-s|\rho(x,y)$ for $s,t \in [0,1]$. A map $\gamma\colon [0,r_0] \to E$ with $\gamma(0)=x$ for some $r_0>0$ and $x\in E$ is called a minimal geodesic from x with speed $c \geq 0$ if $\rho(\gamma(s),\gamma(t)) = c|t-s|$ holds for $s,t \in [0,r_0]$. For a function f on E, we define $|\nabla f|(x)$ as the local Lipschitz constant of f at the point x, i.e.,

$$|\nabla f|(x) = \limsup_{y\to x} \frac{|f(x)-f(y)|}{\rho(x,y)}.$$

Obviously, $|f(x)-f(y)| \leq \rho(x,y)\|\nabla f\|_\infty$; that is, $\|\nabla f\|_\infty$ is the global Lipschitz constant of f. Moreover, a metric space (E,ρ) is called a length space if for every $x \neq y$ and $s \in (0,1)$, there exists a sequence $\{z_n\} \subset E$ such that $\rho(x,z_n) \to s\rho(x,y)$ and $\rho(z_n,y) \to (1-s)\rho(x,y)$ as $n\to\infty$.

1.3.1 Gradient–Entropy and Harnack Inequalities

Returning to [4], the following result provides a link between the gradient–entropy inequality and the Harnack inequality with power. In this subsection we let P be a Markov operator on $\mathscr{B}_b(E)$.

Proposition 1.3.1 *Let (E,ρ) be a geodesic space, and let $\delta_0 \geq 0$ and $\beta \in C((\delta_0,\infty)\times E;[0,\infty))$. The following two statements are equivalent:*

(1) *For every strictly positive $f \in \mathscr{B}_b(E)$,*

$$|\nabla Pf| \leq \delta\{P(f\log f)-(Pf)\log Pf\} + \beta(\delta,\cdot)Pf,\ \ \delta>\delta_0.$$

(2) *For every $p>1$ and $x,y\in E$ such that $\rho(x,y) \leq \frac{p-1}{p\delta_0}$ and for any positive $f \in \mathscr{B}_b(E)$,*

$$(Pf)^p(x) \leq \{Pf^p(y)\} \\ \exp\left[\int_0^1 \frac{p\rho(x,y)}{1+(p-1)s}\beta\Big(\frac{p-1}{\rho(x,y)\{1+(p-1)s\}},\gamma(s)\Big)\mathrm{d}s\right], \tag{1.16}$$

where $\gamma\colon [0,1]\to E$ is a minimal geodesic from x to y with speed $\rho(x,y)$.

Proof. For $p>1$, let $\alpha(s)=1+(p-1)s$. We have $\delta(s) := \frac{p-1}{\alpha(s)\rho(x,y)} > \delta_0$ for $s \in [0,1)$. Then (1) implies that

$$[0,1) \ni s \mapsto \log(Pf^{\alpha(s)})^{\frac{p}{\alpha(s)}}(\gamma(s))$$

is Lipschitz continuous and

$$\begin{aligned}
&\frac{\mathrm{d}}{\mathrm{d}s}\log(Pf^{\alpha(s)})^{\frac{p}{\alpha(s)}}(\gamma(s))\\
&\geq \frac{p(p-1)\{P(f^{\alpha(s)}\log f^{\alpha(s)})-(Pf^{\alpha(s)})\log Pf^{\alpha(s)}\}}{\alpha(s)^2 Pf^{\alpha(s)}}(\gamma(s))\\
&\quad -\frac{p\rho(x,y)|\nabla Pf^{\alpha(s)}|}{\alpha(s)Pf^{\alpha(s)}}(\gamma(s))\\
&\geq -\frac{p\rho(x,y)}{\alpha(s)}\beta\Big(\frac{p-1}{\alpha(s)\rho(x,y)},\ \gamma(s)\Big),\quad s\in[0,1).
\end{aligned}$$

Integrating over $[0,1)$, we obtain (2).

On the other hand, for $z \in E$, let γ be a minimal geodesic from z with $\rho(\gamma(r),z) = r$ for small $r > 0$, and

$$|\nabla f|(z) = \limsup_{r\to 0}\frac{|f(\gamma(r))-f(z)|}{r}.$$

We have either (i) or (ii) as follows:

(i) $|\nabla f|(z) = \limsup_{r\to 0}\frac{f(\gamma(r))-f(z)}{r}$;
(ii) $|\nabla f|(z) = \limsup_{r\to 0}\frac{f(z)-f(\gamma(r))}{r}$.

For $\delta > \delta_0$, let $p = 1+\delta r$. We have $\delta \geq \delta_0(1+\delta r)$ and thus $\rho(\gamma(r),z) = r \leq \frac{p-1}{p\delta_0}$ for small $r > 0$. Applying (2) to $x = \gamma(r)$ and $y = z$, we obtain from (i) that

$$\begin{aligned}
&\delta\{(Pf)\log Pf\}(z) + |\nabla Pf|(z) = \limsup_{r\to 0}\frac{(Pf)^{1+\delta r}(\gamma(r)) - Pf(z)}{r}\\
&\leq \limsup_{r\to 0}\frac{(Pf^{1+\delta r})(z)\exp\Big[\int_0^1\frac{(1+\delta r)r}{1+\delta rs}\beta\Big(\frac{\delta}{1+\delta rs},\gamma(sr)\Big)\mathrm{d}s\Big] - Pf(z)}{r}\\
&= \delta P(f\log f)(z) + \beta(\delta,z)Pf(z).
\end{aligned}$$

Similarly, if (ii) holds, then

$$\begin{aligned}
&|\nabla Pf|(z) - \delta P(f\log f)(z) = \limsup_{r\to 0}\frac{(Pf)(z) - (Pf^{1+\delta r})(\gamma(r))}{r}\\
&\leq \limsup_{r\to 0}\frac{(Pf)(z) - (Pf)^{1+\delta r}(z)\exp\Big[\int_0^1\frac{(1+\delta r)r}{1+\delta rs}\beta\Big(\frac{\delta}{1+\delta rs},\gamma(sr)\Big)\mathrm{d}s\Big]}{r}\\
&= \beta(\delta,z)Pf(z) - \delta\{(Pf)\log Pf\}(z).
\end{aligned}$$

Therefore, (1) holds. □

Similarly, we have the following result on the shift Harnack inequality.

Proposition 1.3.2 *Let E be a Banach space. Let $e \in E$, $\delta_e \in (0,1)$, and $\beta_e \in C((\delta_e,\infty)\times E;[0,\infty))$. Then the following assertions are equivalent:*

(1) *For every positive $f \in C_b^1(E)$,*

$$|P(\nabla_e f)| \le \delta\{P(f\log f) - (Pf)\log Pf\} + \beta_e(\delta,\cdot)Pf, \ \ \delta \ge \delta_e.$$

(2) *For every positive $f \in \mathscr{B}_b(E)$, $r \in (0,\frac{1}{\delta_e})$, and $p \ge \frac{1}{1-r\delta_e}$,*

$$(Pf)^p \le (P\{f^p(re+\cdot)\})\exp\left[\int_0^1 \frac{pr}{1+(p-1)s}\beta_e\Big(\frac{p-1}{r+r(p-1)s}, \cdot + sre\Big)\mathrm{d}s\right].$$

Proof. The proof that (1) implies (2) is completely similar to the first part of the proof in Proposition 1.3.1. To prove that (1) is implied by (2), we let $z, e \in E$ be fixed and assume that $P(\nabla_e f)(z) \ge 0$ (otherwise, simply use $-e$ to replace e). Then (2) with $p = 1+\delta r$ implies that

$$\begin{aligned}
&\delta\{(Pf)\log Pf\}(z) + |P(\nabla_e f)|(z)\\
&= \limsup_{r\to 0}\frac{(P\{f(re+\cdot)\})^{1+\delta r}(z) - Pf(z)}{r}\\
&\le \limsup_{r\to 0}\frac{(Pf^{1+\delta r})(z)\exp\left[\int_0^1 \frac{(1+\delta r)r}{1+\delta rs}\beta_e\Big(\frac{\delta}{1+\delta rs}, z - sre\Big)\mathrm{d}s\right] - Pf(z)}{r}\\
&= \delta P(f\log f)(z) + \beta_e(\delta,z)Pf(z).
\end{aligned}$$

Therefore, (1) holds. □

Finally, we note that the Harnack inequality with a power implies the log-Harnack inequality on a length space.

Proposition 1.3.3 *Let (E,ρ) be a length space. Let $c>0, p_1, p_2 > 1$ be constants. If*

$$(Pf(x))^p \le (Pf^p(y))\exp\left[\frac{pc\rho(x,y)^2}{p-1}\right], \ \ f \in \mathscr{B}_b^+(E),\ x,y \in E, \tag{1.17}$$

holds for $p = p_1, p_2$, then it holds also for $p = p_1p_2$.

Proof. Let

$$s = \frac{p_1-1}{p_1p_2-1}, \ \ 1-s = \frac{p_1(p_2-1)}{p_1p_2-1},$$

and let $\{z_n\} \subset E$ be such that $\rho(x,z_n) \to s\rho(x,y)$ and $\rho(z_n,y) \to (1-s)\rho(x,y)$ as $n \to \infty$. Since (1.17) holds for $p = p_1$ and $p = p_2$, for every $f \in \mathscr{B}_b^+(E)$ we have

$$(Pf(x))^{p_1p_2} \le (Pf^{p_1}(z_n))^{p_2}\exp\left[\frac{p_1p_2c\rho(x,z_n)^2}{p_1-1}\right]$$
$$\le (Pf^{p_1p_2}(y))\exp\left[\frac{p_1p_2c\rho(x,z_n)^2}{p_1-1}+\frac{p_2c\rho(z_n,y)^2}{p_2-1}\right].$$

Letting $n\to\infty$, we arrive at

$$(Pf(x))^{p_1p_2} \le (Pf^{p_1p_2}(y))\exp\left[\frac{p_1p_2cs^2\rho(x,y)^2}{p_1-1}+\frac{p_2c(1-s)^2\rho(x,y)^2}{p_2-1}\right]$$
$$= (Pf^{p_1p_2}(y))\exp\left[\frac{p_1p_2c\rho(x,y)^2}{p_1p_2-1}\right].$$

□

As a consequence of Proposition 1.3.3, (1.17) implies the following log-Harnack inequality (1.18).

Corollary 1.3.4 *Let (E,ρ) be a length space. If (1.17) holds for some $p>1$, then*

$$P(\log f)(x) \le \log Pf(y) + c\rho(x,y)^2,\quad x,y\in E,\ f\ge 1,\ f\in \mathscr{B}_b(E). \tag{1.18}$$

Proof. By Proposition 1.3.3, (1.17) holds for $p^n (n\in\mathbb{N})$ in place of p. So

$$Pf^{p^{-n}}(x) \le (Pf(y))^{p^{-n}}\exp\left[\frac{c\rho(x,y)^2}{p^n-1}\right].$$

Therefore, by the dominated convergence theorem,

$$P(\log f)(x) = \lim_{n\to\infty} P\Big(\frac{f^{p^{-n}}-1}{p^{-n}}\Big)(x)$$
$$\le \lim_{n\to\infty}\left\{\frac{(Pf(y))^{p^{-n}}-1}{p^{-n}} + (Pf(y))^{p^{-n}}\frac{\exp\left[\frac{c\rho(x,y)^2}{p^n-1}\right]-1}{p^{-n}}\right\}$$
$$= \log Pf(y) + c\rho(x,y)^2.$$

□

Completely similarly to Proposition 1.3.3 and Corollary 1.3.4, we have the following result on shift Harnack inequalities.

Theorem 1.3.5 *Let E be a Banach space. Let $c>0$ be a constant.*

(1) *For every $p_1,p_2>1$, if*

$$(Pf(x))^p \le (Pf^p(e+\cdot))(x)\exp\left[\frac{pc|e|^2}{p-1}\right],\quad f\in\mathscr{B}_b^+(E), x,e\in E, \tag{1.19}$$

holds for $p=p_1,p_2$, it holds also for $p=p_1p_2$.

(2) *If (1.19) holds for some* $p > 1$, *then*

$$P(\log f)(x) \le \log Pf(e+\cdot)(x) + c|e|^2, \quad x, e \in E, f \ge 1, f \in \mathscr{B}_b(E). \tag{1.20}$$

1.3.2 From Gradient–Gradient to Harnack Inequalities

In this subsection we consider the diffusion semigroup P_t with generator $(L, \mathscr{D}(L))$ on a geodesic space (E, ρ) in the following sense: there exists a subclass $\mathscr{A}_0 \subset \mathscr{D}(L)$ of $\mathscr{B}_b(E)$ such that for every $f \in \mathscr{A}_0$ and $\varphi \in C^\infty([\inf f, \sup f])$ one has $P_t f, \varphi \circ f \in \mathscr{A}_0$ and

$$\frac{\mathrm{d}}{\mathrm{d}t} P_t f = P_t L f = L P_t f, \quad L\varphi \circ f = \varphi' \circ f L f + \varphi'' \circ f \Gamma(f), \quad t \ge 0, \tag{1.21}$$

where $\Gamma(f) \ge \lambda |\nabla f|^2$ for some constant $\lambda > 0$. A typical example is a nonexplosive elliptic diffusion process on a differential manifold E. In this case, we take ρ to be the intrinsic metric induced by the square field of the diffusion, and let

$$\mathscr{A}_0 = \left\{ P_t f : t \ge 0, f \in C^\infty, \mathrm{d}f \text{ has compact support} \right\}.$$

Theorem 1.3.6 *Assume that (1.21) holds with* $\Gamma(f) \ge \lambda |\nabla f|^2$ *for some constant* $\lambda > 0$. *Let* ξ *be a positive measurable function on* $[0, \infty)$, *and let* $g \in C^1([0, t])$ *be increasing with* $g(0) = 0$ *and* $g(t) = 1$.

(1) *If*

$$|\nabla P_t f|^2 \le \xi(t)^2 P_t |\nabla f|^2, \quad f \in \mathscr{A}_0, \ t \ge 0, \tag{1.22}$$

then

$$P_t f(y) \le \log P_t \mathrm{e}^f(x) + \frac{\rho(x,y)^2}{4\lambda} \int_0^t |g'(s)\xi(s)|^2 \mathrm{d}s, \quad t > 0, f \in \mathscr{A}_0. \tag{1.23}$$

(2) *If*

$$|\nabla P_t f| \le \xi(t) P_t |\nabla f|, \quad f \in \mathscr{A}_0, \ t \ge 0, \tag{1.24}$$

then

$$(P_t f)^p(x) \le (P_t f^p(y)) \exp\left[\frac{p\rho(x,y)^2}{4(p-1)\lambda} \int_0^t \left|\xi(s) g'(s)\right|^2 \mathrm{d}s \right] \tag{1.25}$$

holds for $t > 0$ *and nonnegative* $f \in \mathscr{A}_0$.

Proof. Let $\gamma : [0,1] \to E$ be a minimal geodesic from x to y with constant speed $\rho(x,y)$.

(1) By (1.21) and (1.22), we have

$$
\begin{aligned}
&\frac{\mathrm{d}}{\mathrm{d}s}P_s \log P_{t-s}\mathrm{e}^f(\gamma\circ g(s)) \\
&\le \Big\{-\lambda P_s|\nabla \log P_{t-s}\mathrm{e}^f|^2 + \rho(x,y)g'(s)\cdot|\nabla P_s \log P_{t-s}\mathrm{e}^f|\Big\}(\gamma\circ g(s)) \\
&\le \Big\{-\lambda P_s|\nabla \log P_{t-s}\mathrm{e}^f|^2 + \rho(x,y)g'(s)\xi(s)\sqrt{P_s|\nabla \log P_{t-s}\mathrm{e}^f|^2}\Big\}(\gamma\circ g(s)) \\
&\le \frac{\rho(x,y)^2\xi(s)^2|g'(s)|^2}{4\lambda}.
\end{aligned}
$$

Integrating over $[0,t]$, we obtain (1.23).

(2) Similarly, by (1.21) and (1.24), we obtain

$$
\begin{aligned}
&\frac{\mathrm{d}}{\mathrm{d}s}P_s(P_{t-s}f)^p(\gamma\circ g(s)) \\
&\ge P_s\{p(p-1)\lambda(P_{t-s}f)^{p-2}|\nabla P_{t-s}f|^2\}(\gamma\circ g(s)) \\
&\quad -\rho(x,y)g'(s)\cdot\big|\nabla P_s(P_{t-s}f)^p(\gamma\circ g(s))\big| \\
&\ge pP_s\Big\{(P_{t-s}f)^p\Big((p-1)\lambda\frac{|\nabla P_{t-s}f|^2}{(P_{t-s}f)^2} - \rho(x,y)g'(s)\xi(s)\cdot\frac{|\nabla P_{t-s}f|}{P_{t-s}f}\Big)\Big\}(\gamma\circ g(s)) \\
&\ge -\frac{p\rho(x,y)^2\xi(s)^2|g'(s)|^2}{4(p-1)\lambda}P_s(P_{t-s}f)^p(\gamma\circ g(s)), \quad s\in[0,t]. \qquad (1.26)
\end{aligned}
$$

This implies (1.25). □

We remark that Theorem 1.3.6 (1) and (2) go back to [43] and [50] respectively, where log-Harnack and Harnack inequalities were established for semilinear SPDEs and diffusion semigroups on manifolds. The arguments have been applied in [1, 2, 23, 41, 69] for some other models.

1.3.3 L^2 Gradient and Harnack Inequalities

The following result is taken from [60].

Theorem 1.3.7 *For every constant $C>0$, the gradient-L^2 estimate*

$$|\nabla Pf|^2 \le C^2 Pf^2, \quad f\in\mathscr{B}_b(E), \qquad (1.27)$$

is equivalent to the Harnack-type inequality

$$Pf(z') \le Pf(z) + C\rho(z,z')\sqrt{Pf^2(z')}, \qquad (1.28)$$

which holds for all $z,z'\in E$, $f\in\mathscr{B}_b^+(E)$.

Proof. **(1.27) ⇒ (1.28)**. Let $\gamma:[0,1]\to E$ be a minimal geodesic such that $\gamma(0)=z$, $\gamma(1)=z'$ with constant speed $\rho(z,z')$. By (1.27), for every positive $f\in\mathscr{B}_b(E)$ and constant $r>0$, we have

$$\begin{aligned}
&\frac{\mathrm{d}}{\mathrm{d}s}P\Big(\frac{f}{1+rsf}\Big)(\gamma(s))\\
&\le -rP\Big(\frac{f^2}{(1+rsf)^2}\Big)(\gamma(s))+C\rho(z,z')\sqrt{P\Big(\frac{f}{1+rsf}\Big)^2(\gamma(s))} \qquad (1.29)\\
&\le \frac{C^2\rho(z,z')^2}{4r}. \qquad (1.30)
\end{aligned}$$

So

$$P\Big(\frac{f}{1+rf}\Big)(z')\le Pf(z)+\frac{C^2\rho(z,z')^2}{4r}.$$

Combining this with the fact that

$$\frac{f}{1+rf}=f-\frac{rf^2}{1+rf}\ge f-rf^2,$$

we obtain

$$Pf(z')\le Pf(z)+\frac{C^2\rho(z,z')^2}{4r}+rPf^2(z').$$

Minimizing the right-hand side in $r>0$, we prove (1.28).

(1.28) ⇒ (1.27). By (1.28), we have

$$|Pf(z)-Pf(z')|\le C\rho(z,z')\|f\|_\infty,\ \ f\in\mathscr{B}_b(E).$$

So Pf is Lipschitz continuous for every $f\in\mathscr{B}_b(E)$. Let $z\in E$ and let $\gamma:[0,1]\to E$ be a minimal geodesic such that $\gamma(0)=z$, $\rho(\gamma(0),\gamma(s))=s$, and

$$\limsup_{s\to0}\frac{|Pf(\gamma(s))-Pf(z)|}{s}=|\nabla Pf|(z).$$

Then it follows from (1.28) that

$$\begin{aligned}
|\nabla Pf|(z)&=\limsup_{s\to0}\frac{|Pf(\gamma(s))-Pf(\gamma(0))|}{s}\\
&\le C\lim_{s\to0}\max\Big\{\sqrt{Pf^2(\gamma(s))},\sqrt{Pf^2(z)}\Big\}=C\sqrt{Pf^2(z)}.
\end{aligned}$$

Therefore, (1.27) holds. □

Next, according to [5], we show that the log-Harnack inequality implies the gradient-L^2 estimate.

Proposition 1.3.8 *Let $x\in E$ be fixed. If there exists a positive function $G\in C(E^2)$ such that the log-Harnack inequality*

$$P\log f(y)\le\log Pf(x)+G(x,y)\rho(x,y)^2,\ \ f>0,\ f\in\mathscr{B}_b(E), \qquad (1.31)$$

holds for small $\rho(x,y)$, *then*

$$|\nabla Pf|^2(x) \le 2G(x,x)\big\{Pf^2(x) - (Pf)^2(x)\big\}, \ f \in \mathscr{B}_b(E). \tag{1.32}$$

Proof. Let $f \in \mathscr{B}_b(E)$. According to the proof of Theorem 1.4.1(1) below, (1.31) for small $\rho(x,y)$ implies that Pf is continuous at x. Let $\{x_n\}_{n\ge1}$ be a sequence converging to x, and define $\varepsilon_n = \rho(x_n,x)$. For every positive constant $c > 0$, we apply (1.31) to $c\varepsilon_n f + 1$ in place of f, so that for large enough n,

$$P\log(c\varepsilon_n f + 1)(x_n) \le \log\big\{P(c\varepsilon_n f + 1)(x)\big\} + G(x,x_n)\varepsilon_n^2. \tag{1.33}$$

Noting that for large n (or for small ε_n) we have

$$\begin{aligned} P\log(c\varepsilon_n f+1)(x_n) &= P\Big(c\varepsilon_n f - \frac{1}{2}(c\varepsilon_n)^2 f^2\Big)(x_n) + o(\varepsilon_n^2) \\ &= c\varepsilon_n Pf(x) + c\varepsilon_n^2 \frac{Pf(x_n)-Pf(x)}{\rho(x_n,x)} - \frac{1}{2}(c\varepsilon_n)^2 Pf^2(x) + o(\varepsilon_n^2), \\ \log\big\{P(c\varepsilon_n f+1)(x)\big\} &= c\varepsilon_n Pf(x) - \frac{1}{2}(c\varepsilon_n)^2 (Pf)^2(x) + o(\varepsilon_n^2), \end{aligned}$$

it follows from (1.33) that

$$c\limsup_{n\to\infty} \frac{Pf(x_n)-Pf(x)}{\rho(x_n,x)} \le \frac{c^2}{2}\{Pf^2(x) - (Pf)^2(x)\} + G(x,x), \ c>0.$$

Exchanging the positions of x_n and x, we obtain

$$c\limsup_{n\to\infty} \frac{Pf(x)-Pf(x_n)}{\rho(x_n,x)} \le \frac{c^2}{2}\{Pf^2(x) - (Pf)^2(x)\} + G(x,x), \ c>0.$$

Since the sequence $x_n \to x$ is arbitrary, these imply

$$|\nabla Pf|(x) \le \frac{c}{2}\{Pf^2(x) - (Pf)^2(x)\} + \frac{G(x,x)}{c}, \ c>0.$$

By minimizing the upper bound in $c > 0$, we prove (1.32). □

Correspondingly, we have the following result due to [61] concerning the shift Harnack inequalities.

Proposition 1.3.9 *Let E be a Banach space and $C \ge 0$ a constant.*

(1) *For every $e \in E$, $|P(\nabla_e f)|^2 \le CPf^2$, $f \in C_b^1(E)$, $f \ge 0$, is equivalent to*

$$Pf \le P\{f(re+\cdot)\} + |r|\sqrt{CPf^2}, \ r \in \mathbb{R}, f \in \mathscr{B}_b^+(E).$$

(2) *Let $x,e \in E$ be fixed. If there exists a positive function G on E^2 continuous in the first variable such that the shift log-Harnack inequality*

$$P(\log f)(x) \le \log Pf(re+\cdot)(x) + r^2 G(re,x)|e|^2, \ f \in \mathscr{B}_b(E),$$

holds for small $r > 0$, then

$$|P(\nabla_e f)(x)|^2 \le 2G(0,x)\{Pf^2(x) - (Pf)^2(x)\}, \quad f \in \mathscr{B}_b(E).$$

1.4 Applications of Harnack and Shift Harnack Inequalities

In this section we collect some applications of Harnack and shift-Harnack inequalities for heat kernel estimates, invariant probability measures, and cost-entropy inequalities. These results can be then applied to various specific models for which Harnack and shift Harnack inequalities are derived in the other three chapters. See also [6, 51, 52, 56, 59] for links of Harnack inequalities to curvature conditions, second fundamental forms, and optimal transportation.

Definition 1.4.1 *Let μ be a probability measure on $(E, \mathscr{B})$, and let P be a bounded linear operator on $\mathscr{B}_b(E)$.*

(*i*) *The measure μ is called a* quasi-invariant probability measure *of P if μP is absolutely continuous with respect to μ, where $(\mu P)(A) := \mu(P1_A)$, $A \in \mathscr{B}$. If $\mu P = \mu$, then μ is called an* invariant probability measure *of P.*
(*ii*) *A measurable function $\mathbf{p}$ on E^2 is called the* kernel *or* density *of P with respect to μ if*

$$Pf = \int_E \mathbf{p}(\cdot, y) f(y) \mu(\mathrm{d}y), \quad f \in \mathscr{B}_b(E).$$

(*iii*) *Let E be a topological space. The operator P is called a* Feller *operator if $PC_b(E) \subset C_b(E)$, while it is called a* strong Feller *operator if $P\mathscr{B}_b(E) \subset C_b(E)$.*

Throughout this section, we assume that E is a topological space with Borel σ-field $\mathscr{B}$ and that P is a Markov operator given by (1.1) for some transition probability measure $\{\mu_x\}_{x \in E}$.

1.4.1 Applications of the Harnack Inequality

We will consider applications of the Harnack-type inequality (1.2), where $\Phi \in C([0,\infty))$ is nonnegative and strictly increasing, and Ψ is a measurable nonnegative function on E^2. Results presented in this subsection are taken or modified from [3, 4, 10, 19, 41, 43, 54, 56, 68].

Theorem 1.4.1 *Let μ be a quasi-invariant probability measure of P. Let $\Phi \in C^1([0,\infty))$ be an increasing function with $\Phi'(1) > 0$ and $\Phi(\infty) := \lim_{r \to \infty} \Phi(r) = \infty$ such that (1.2) holds.*

(1) *If* $\lim_{y\to x}\{\Psi(x,y)+\Psi(y,x)\}=0$ *holds for all* $x\in E$, *then* P *is strong Feller.*
(2) P *has a kernel* $\mathbf{p}$ *with respect to* μ, *so that every invariant probability measure of* P *is absolutely continuous with respect to* μ.
(3) P *has at most one invariant probability measure, and if it has one, the kernel of* P *with respect to the invariant probability measure is strictly positive.*
(4) *The kernel* $\mathbf{p}$ *of* P *with respect to* μ *satisfies*

$$\int_E \mathbf{p}(x,\cdot)\Phi^{-1}\Big(\frac{\mathbf{p}(x,\cdot)}{\mathbf{p}(y,\cdot)}\Big)\mathrm{d}\mu \le \Phi^{-1}(\mathrm{e}^{\Psi(x,y)}),\ \ x,y\in E,$$

where $\Phi^{-1}(\infty):=\infty$ *by convention.*
(5) *If* $r\Phi^{-1}(r)$ *is convex for* $r\ge 0$, *then the kernel* $\mathbf{p}$ *of* P *with respect to* μ *satisfies*

$$\int_E \mathbf{p}(x,\cdot)\mathbf{p}(y,\cdot)\mathrm{d}\mu \ge \mathrm{e}^{-\Psi(x,y)},\ \ x,y\in E.$$

(6) *If* μ *is an invariant probability measure of* P, *then*

$$\sup_{f\in\mathscr{B}_b^+(E),\mu(\Phi(f))\le 1}\Phi(Pf(x))\le \frac{1}{\int_E \mathrm{e}^{-\Psi(x,y)}\mu(\mathrm{d}y)},\ \ x\in E.$$

Proof. Since (6) is obvious, below we prove (1)–(5).

(1) Let $f\in\mathscr{B}_b(E)$ be positive. Applying (1.2) to $1+\varepsilon f$ in place of f for $\varepsilon>0$, we have

$$\Phi(1+\varepsilon Pf(x))\le \{P\Phi(1+\varepsilon f)(y)\}\mathrm{e}^{\Psi(x,y)},\ \ x,y\in E,\varepsilon>0.$$

By a Taylor expansion, this implies

$$\Phi(1)+\varepsilon\Phi'(1)Pf(x)+\mathrm{o}(\varepsilon)\le\{\Phi(1)+\varepsilon\Phi'(1)Pf(y)+\mathrm{o}(\varepsilon)\}\mathrm{e}^{\Psi(x,y)} \quad (1.34)$$

for small $\varepsilon>0$. Letting $y\to x$, we obtain

$$\varepsilon Pf(x)\le\varepsilon\liminf_{y\to x}Pf(y)+\mathrm{o}(\varepsilon).$$

Thus, $Pf(x)\le\liminf_{y\to x}Pf(y)$ holds for all $x\in E$. On the other hand, letting $x\to y$ in (1.34) gives $Pf(y)\ge\limsup_{x\to y}Pf(x)$ for every $y\in E$. Therefore, Pf is continuous.

(2) To prove the existence of a kernel, it suffices to prove that for every $A\in\mathscr{B}$ with $\mu(A)=0$, we have $P1_A\equiv 0$. Applying (1.2) to $f=1+n1_A$, we obtain

$$\Phi(1+nP1_A(x))\int_E \mathrm{e}^{-\Psi(x,y)}\mu(\mathrm{d}y)\le\int_E\Phi(1+n1_A)(y)(\mu P)(\mathrm{d}y),\ \ n\ge 1. \quad (1.35)$$

Since $\mu(A)=0$ and μ is quasi-invariant for P, we have $1_A=0$, μP-a.s. So it follows from (1.35) that

$$\Phi(1+nP1_A(x)) \le \frac{\Phi(1)}{\int_E e^{-\Psi(x,y)}\mu(dy)} < \infty, \quad x \in E, n \ge 1.$$

Since $\Phi(1+n) \to \infty$ as $n \to \infty$, this implies that $P1_A(x) = 0$ for all $x \in E$. Now, for any invariant probability measure μ_0 of P, if $\mu(A) = 0$, then $P1_A \equiv 0$ implies that $\mu_0(A) = \mu_0(P1_A) = 0$. Therefore, μ_0 is absolutely continuous with respect to μ.

(3) We first prove that the kernel of P with respect to an invariant probability measure μ_0 is strictly positive. To this end, it suffices to show that for every $x \in E$ and $A \in \mathscr{B}$, $P1_A(x) = 0$ implies that $\mu_0(A) = 0$. Since $P1_A(x) = 0$, applying (1.2) to $f = 1 + n1_A$, we obtain

$$\Phi(1+nP1_A(y)) \le \{P\Phi(1+n1_A)(x)\}e^{\Psi(y,x)} = \Phi(1)e^{\Psi(y,x)}, \quad y \in E, n \ge 1.$$

Letting $n \to \infty$, we conclude that $P1_A \equiv 0$ and hence $\mu_0(A) = \mu_0(P1_A) = 0$. Next, let μ_1 be another invariant probability measure of P. By (2), we have $d\mu_1 = f d\mu_0$ for some probability density function f. We aim to prove that $f = 1$, μ_0-a.e. Let $\mathbf{p}(x,y) > 0$ be the kernel of P with respect to μ_0, and let $P^*(x,dy) = \mathbf{p}(y,x)\mu_0(dy)$. Then

$$P^*g = \int_E g(y)P^*(\cdot,dy), \quad g \in \mathscr{B}_b(E),$$

is the adjoint operator of P with respect to μ_0. Since μ_0 is P-invariant, we have

$$\int_E gP^*1\,d\mu_0 = \int_E Pg\,d\mu_0 = \int_E g\,d\mu_0, \quad g \in \mathscr{B}_b(E).$$

This implies that $P^*1 = 1$, μ_0-a.e. Thus, for μ_0-a.e. $x \in E$, the measure $P^*(x,\cdot)$ is a probability measure. On the other hand, since μ_1 is P-invariant, we have

$$\int_E (P^*f)g\,d\mu_0 = \int_E fPg\,d\mu_0 = \int_E Pg\,d\mu_1 = \int_E g\,d\mu_1 = \int_E fg\,d\mu_0, \quad g \in \mathscr{B}_b(E).$$

This implies that $P^*f = f$, μ_0-a.e. Therefore,

$$\int_E P^*\frac{1}{f+1}\,d\mu_0 = \int_E \frac{1}{f+1}\,d\mu_0 = \int_E \frac{1}{P^*f+1}\,d\mu_0.$$

When $P^*(x,\cdot)$ is a probability measure, by Jensen's inequality one has $P^*\frac{1}{1+f}(x) \ge \frac{1}{P^*f+1}(x)$, and the equation holds if and only if f is constant $P^*(x,\cdot)$-a.s. Hence, f is constant $P^*(x,\cdot)$-a.s. for μ_0-a.e. x. Since $\mathbf{p}(x,y) > 0$ for every $y \in E$ such that μ_0 is absolutely continuous with respect to $P^*(x,\cdot)$ for every $x \in E$, we conclude that f is constant μ_0-a.s. Therefore, $f = 1$ μ_0-a.s., since f is a probability density function.

(4) Applying (1.2) to

$$f = n \wedge \Phi^{-1}\Big(\frac{\mathbf{p}(x,\cdot)}{\mathbf{p}(y,\cdot)}\Big)$$

and letting $n \to \infty$, we obtain the desired inequality.

(5) Let $r\Phi^{-1}(r)$ be convex for $r \geq 0$. By Jensen's inequality, we have

$$\int_E \mathbf{p}(x,\cdot)\Phi^{-1}(\mathbf{p}(x,\cdot))\mathrm{d}\mu \geq \Phi^{-1}(1).$$

So, applying (1.2) to

$$f = n \wedge \Phi^{-1}(\mathbf{p}(x,\cdot))$$

and letting $n \to \infty$, we obtain

$$\int_E \mathbf{p}(x,\cdot)\mathbf{p}(y,\cdot)\mathrm{d}\mu \geq \mathrm{e}^{-\Psi(x,y)}\Phi\left(\int_E \mathbf{p}(x,\cdot)\Phi^{-1}(\mathbf{p}(x,\cdot))\mathrm{d}\mu\right) \geq \mathrm{e}^{-\Psi(x,y)}.$$

□

Next, we present some additional applications of the Harnack inequality with power and the log-Harnack inequality.

Theorem 1.4.2 *Let Ψ be a positive function on $E \times E$.*

(1) *(1.3) holds if and only if μ_x and μ_y in (1.1) are equivalent and $\mathbf{p}_{x,y} := \frac{\mathrm{d}\mu_x}{\mathrm{d}\mu_y}$ satisfies*

$$P\{\mathbf{p}_{x,y}^{\frac{1}{p-1}}\}(x) \leq \exp\left[\frac{\Psi(x,y)}{p-1}\right], \quad x,y \in E. \tag{1.36}$$

(2) *(1.4) holds if and only if μ_x and μ_y are equivalent and $\mathbf{p}_{x,y}$ satisfies*

$$P\{\log \mathbf{p}_{x,y}\}(x) \leq \Psi(x,y), \quad x,y \in E. \tag{1.37}$$

(3) *If (1.4) holds, then for a P-invariant probability measure μ, the entropy–cost inequality*

$$\mu((P^*f)\log P^*f) \leq W_1^\Psi(f\mu,\mu), \;\; f \geq 0, \mu(f) = 1,$$

holds for P^*, the adjoint operator of P in $L^2(\mu)$, where W_1^Ψ is the L^1 transportation cost induced by the cost function Ψ, i.e., for any two probability measures μ_1,μ_2 on E,

$$W_1^\Psi(\mu_1,\mu_2) := \inf_{\pi\in\mathscr{C}(\mu_1,\mu_2)} \int_{E\times E} \Psi \mathrm{d}\pi,$$

where $\mathscr{C}(\mu_1,\mu_2)$ is the set of all couplings of μ_1,μ_2.

Proof. It is easy to see that each of (1.3) and (1.4) implies the equivalence of μ_x and μ_y. Below, we prove the desired inequalities on $\mathbf{p}_{x,y}$.

(1) Applying (1.3) to $f_n(z) := \{n \wedge \mathbf{p}_{x,y}(z)\}^{\frac{1}{p-1}}, n \geq 1$, we obtain

$$\begin{aligned}(Pf_n(x))^p &\leq \mathrm{e}^{\Psi(x,y)}Pf_n^p(y) = \mathrm{e}^{\Psi(x,y)}\int_E \{n \wedge \mathbf{p}_{x,y}(z)\}^{\frac{p}{p-1}}\mu_y(\mathrm{d}z)\\ &\leq \mathrm{e}^{\Psi(x,y)}\int_E \{n \wedge \mathbf{p}_{x,y}(z)\}^{\frac{1}{p-1}}\mu_x(\mathrm{d}z) = \mathrm{e}^{\Psi(x,y)}Pf_n(x).\end{aligned}$$

Thus,

$$P\{\mathbf{p}_{x,y}^{\frac{1}{p-1}}\}(x) = \lim_{n\to\infty} Pf_n(x) \le \exp\left[\frac{\Psi(x,y)}{p-1}\right].$$

On the other hand, if (1.36) holds, then for every $f \in \mathscr{B}_b^+(E)$, by Hölder's inequality we have

$$\begin{aligned}
Pf(x) &= \int_E \mathbf{p}_{x,y}(z) f(z)\mu_y(\mathrm{d}z) \\
&\le (Pf^p(y))^{\frac{1}{p}} \left(\int_E \mathbf{p}_{x,y}(z)^{\frac{p}{p-1}}\mu_y(\mathrm{d}z)\right)^{\frac{p-1}{p}} \\
&= (Pf^p(y))^{\frac{1}{p}} (P\mathbf{p}_{x,y}^{\frac{1}{p-1}}(x))^{\frac{p-1}{p}} \\
&\le (Pf^p(y))^{\frac{1}{p}} \mathrm{e}^{\frac{\Psi(x,y)}{p}}.
\end{aligned}$$

Therefore, (1.3) holds.

(2) We shall use the following Young's inequality: for a probability measure ν on E, if $g_1, g_2 \ge 0$ with $\nu(g_1) = 1$, then

$$\nu(g_1 g_2) \le \nu(g_1 \log g_1) + \log \nu(\mathrm{e}^{g_2}).$$

For $f \ge 1$, applying the above inequality for $g_1 = \mathbf{p}_{x,y}$, $g_2 = \log f$, and $\nu = \mu_y$, we obtain

$$P(\log f)(x) = \int_E \{\mathbf{p}_{x,y}(z)\log f(z)\}\mu_y(\mathrm{d}z) \le P(\log \mathbf{p}_{x,y})(x) + \log Pf(y).$$

So (1.37) implies (1.4). On the other hand, applying (1.4) to $f_n = 1 + n\mathbf{p}_{x,y}$, we arrive at

$$\begin{aligned}
&P\{\log \mathbf{p}_{x,y}\}(x) \le P(\log f_n)(x) - \log n \\
&\le \log Pf_n(y) - \log n + \Psi(x,y) = \log\frac{n+1}{n} + \Psi(x,y).
\end{aligned}$$

Therefore, by letting $n \to \infty$, we obtain (1.37).

(3) Let $\Pi \in \mathscr{C}(f\mu, \mu)$. Applying (1.4) to $P^* f$ in place of f and integrating with respect to Π, we obtain

$$\begin{aligned}
&\mu((P^*f)\log P^*f) = \int_{E\times E} P\log P^* f(x)\Pi(\mathrm{d}x,\mathrm{d}y) \\
&\le \int_{E\times E} \log PP^* f(y)\Pi(\mathrm{d}x,\mathrm{d}y) + \Pi(\Psi) = \mu(\log PP^* f) + \Pi(\Psi) \\
&\le \log \mu(PP^* f) + \Pi(\Psi) = \Pi(\Psi),
\end{aligned}$$

where in the last two steps we have used Jensen's inequality and that μ is PP^*-invariant. This completes the proof. □

1.4.2 Applications of the Shift Harnack Inequality

This subsection is based on [61]. Let P be a Markov operator on a Banach space E. Let $\Phi : [0,\infty) \to [0,\infty)$ be a strictly increasing and convex continuous function. Consider the shift Harnack inequality

$$\Phi(Pf(x)) \le P\{\Phi\circ f(e+\cdot)\}(x)\mathrm{e}^{C_\Phi(x,e)},\ \ f\in \mathscr{B}_b^+(E), \tag{1.38}$$

for some $x,e \in E$ and constant $C_\Phi(x,e)\ge 0$. Obviously, if $\Phi(r)=r^p$ for some $p>1$, then this inequality reduces to the shift Harnack inequality with power p, while when $\Phi(r)=\mathrm{e}^r$, it becomes the shift log-Harnack inequality.

Theorem 1.4.3 *Let $E=\mathbb{R}^d$ and assume that (1.38) holds. Then*

$$\sup_{f\in\mathscr{B}_b^+(\mathbb{R}^d),\int_{\mathbb{R}^d}\Phi\circ f(x)\mathrm{d}x\le 1}\Phi(Pf)(x)\le \frac{1}{\int_{\mathbb{R}^d}\mathrm{e}^{-C_\Phi(x,e)}\mathrm{d}e},\ \ x\in\mathbb{R}^d. \tag{1.39}$$

Consequently:

(1) *If $\Phi(0)=0$, then P has transition density $\mathbf{p}(x,y)$ with respect to the Lebesgue measure such that*

$$\int_{\mathbb{R}^d}\mathbf{p}(x,y)\Phi^{-1}(\mathbf{p}(x,y))\mathrm{d}y\le\Phi^{-1}\bigg(\frac{1}{\int_{\mathbb{R}^d}\mathrm{e}^{-C_\Phi(x,e)}\mathrm{d}e}\bigg). \tag{1.40}$$

(2) *If $\Phi(r)=r^p$ for some $p>1$, then*

$$\int_{\mathbb{R}^d}\mathbf{p}(x,y)^{\frac{p}{p-1}}\mathrm{d}y\le\frac{1}{(\int_{\mathbb{R}^d}\mathrm{e}^{-C_\Phi(x,e)}\mathrm{d}e)^{\frac{1}{p-1}}}. \tag{1.41}$$

Proof. Let $f\in\mathscr{B}_b^+(\mathbb{R}^d)$ be such that $\int_{\mathbb{R}^d}\Phi(f)(x)\mathrm{d}x\le 1$. By (1.38), we have

$$\Phi(Pf)(x)\mathrm{e}^{-C_\Phi(x,e)}\le P(\Phi\circ f(e+\cdot))(x)=\int_{\mathbb{R}^d}\Phi\circ f(y+e)\mu_x(\mathrm{d}y).$$

Integrating both sides with respect to $\mathrm{d}e$ and noting that $\int_{\mathbb{R}^d}\Phi\circ f(y+e)\mathrm{d}e=\int_{\mathbb{R}^d}\Phi\circ f(e)\mathrm{d}e\le 1$, we obtain

$$\Phi(Pf)(x)\int_{\mathbb{R}^d}\mathrm{e}^{-C_\Phi(x,e)}\mathrm{d}e\le 1.$$

This implies (1.39). When $\Phi(0)=0$, (1.39) implies that

$$\sup_{f\in\mathscr{B}_b^+(\mathbb{R}^d),\int_{\mathbb{R}^d}\Phi\circ f(x)\mathrm{d}x\le 1}Pf(x)\le\Phi^{-1}\bigg(\frac{1}{\int_{\mathbb{R}^d}\mathrm{e}^{-C_\Phi(x,e)}\mathrm{d}e}\bigg)<\infty, \tag{1.42}$$

since by the strictly increasing and convex properties, we have $\Phi(r)\uparrow\infty$ and $r\uparrow\infty$. Now, for any Lebesgue-null set A (i.e., set of Lebesgue measure zero), taking $f_n=n1_A$, we obtain from $\Phi(0)=0$ that

$$\int_{\mathbb{R}^d} \Phi \circ f_n(x)\mathrm{d}x = 0 \le 1.$$

Therefore, applying (1.42) to $f = f_n$, we obtain

$$\mu_x(A) = P1_A(x) \le \frac{1}{n}\Phi^{-1}\left(\frac{1}{\int_{\mathbb{R}^d} \mathrm{e}^{-C_\Phi(x,e)}\mathrm{d}e}\right),$$

which goes to zero as $n \to \infty$. Thus, μ_x is absolutely continuous with respect to Lebesgue measure, so that the density function $\mathbf{p}(x,y)$ exists, and (1.40) follows from (1.39) by taking $f(y) = \Phi^{-1}(\mathbf{p}(x,y))$.

Finally, let $\Phi(r) = r^p$ for some $p > 1$. For fixed x, let

$$f_n(y) = \frac{\{n \wedge \mathbf{p}(x,y)\}^{\frac{1}{p-1}}}{\left(\int_{\mathbb{R}^d} \{n \wedge \mathbf{p}(x,y)\}^{\frac{p}{p-1}}\mathrm{d}y\right)^{\frac{1}{p}}}, \quad n \ge 1.$$

It is easy to see that $\int_{\mathbb{R}^d} f_n^p(y)\mathrm{d}y = 1$. Then it follows from (1.39) with $\Phi(r) = r^p$ that

$$\int_{\mathbb{R}^d} \{n \wedge \mathbf{p}(x,y)\}^{\frac{p}{p-1}}\mathrm{d}y \le (Pf_n(x))^{\frac{p}{p-1}} \le \frac{1}{\left(\int_{\mathbb{R}^d} \mathrm{e}^{-C_\Phi(x,e)}\mathrm{d}e\right)^{\frac{1}{p-1}}}.$$

Then (1.41) follows by letting $n \to \infty$. □

Finally, we consider applications of the shift Harnack inequality to distribution properties of the underlying transition probability.

Theorem 1.4.4 *Let (1.38) hold for some $x, e \in E$, finite $C_\Phi(x,e)$, and some strictly increasing, convex, continuous function Φ with $\Phi(0) = 0$. Then μ_x is absolutely continuous with respect to $\mu_x(\cdot - e)$, and the density $\mathbf{p}(x,e;y) := \frac{\mu_x(\mathrm{d}y)}{\mu_x(\mathrm{d}y - e)}$ satisfies*

$$\int_E \Phi^{-1}(\mathbf{p}(x,e;y))\mu_x(\mathrm{d}y) \le \Phi^{-1}\left(\mathrm{e}^{C_\Phi(x,e)}\right).$$

Proof. For a $\mu_x(\cdot - e)$-null set A, let $f = 1_A$. Then (1.38) implies that $\Phi(\mu_x(A)) \le 0$; hence $\mu_x(A) = 0$, since $\Phi(r) > 0$ for $r > 0$. Therefore, μ_x is absolutely continuous with respect to $\mu_x(\cdot - e)$. Next, let $f(y) = \Phi^{-1}(n \wedge \mathbf{p}(x,e;y))$. Noting that

$$P\{\Phi \circ f(e + \cdot)\}(x) = \int_E \frac{\Phi \circ f(y)}{\mathbf{p}(x,e;y)}\mu_x(\mathrm{d}y) = \int_E \frac{n \wedge \mathbf{p}(x,e;y)}{\mathbf{p}(x,e;y)}\mu_x(\mathrm{d}y),$$

by applying (1.38) and letting $n \to \infty$, we finish the proof. □

Chapter 2
Nonlinear Monotone Stochastic Partial Differential Equations

2.1 Solutions of Monotone Stochastic Equations

Let $\mathbb{V} \subset \mathbb{H} \subset \mathbb{V}^*$ be a Gelfand triple, i.e., $(\mathbb{H}, \langle\cdot,\cdot\rangle, |\cdot|)$ is a separable Hilbert space, $\mathbb{V}$ is a reflexive Banach space continuously and densely embedded into $\mathbb{H}$, and $\mathbb{V}^*$ is the duality of $\mathbb{V}$ with respect to $\mathbb{H}$. Let ${}_{\mathbb{V}^*}\langle\cdot,\cdot\rangle_{\mathbb{V}}$ be the dualization between $\mathbb{V}$ and $\mathbb{V}^*$. We have ${}_{\mathbb{V}^*}\langle u,v\rangle_{\mathbb{V}} = \langle u,v\rangle$ for $u \in \mathbb{H}$ and $v \in \mathbb{V}$. Let $\mathscr{L}(\mathbb{H})$ (respectively $\mathscr{L}_b(\mathbb{H})$, $\mathscr{L}_{HS}(\mathbb{H})$) be the set of all densely defined (respectively bounded, Hilbert–Schmidt) linear operators on $\mathbb{H}$. Let $\|\cdot\|$ and $\|\cdot\|_{HS}$ denote the operator norm and the Hilbert–Schmidt norm respectively.

Let $W = (W(t))_{t\geq 0}$ be the cylindrical Brownian motion on $\mathbb{H}$ (see the beginning of Sect. 1.2) with respect to a complete filtered probability space $(\Omega, \mathscr{F}, \{\mathscr{F}_t\}_{t\geq 0}, \mathbb{P})$. Consider the following stochastic equation:

$$\mathrm{d}X(t) = b(t,X(t))\mathrm{d}t + \sigma(t,X(t))\mathrm{d}W(t), \tag{2.1}$$

where

$$b : [0,\infty) \times \mathbb{V} \times \Omega \to \mathbb{V}^*, \quad \sigma : [0,\infty) \times \mathbb{V} \times \Omega \to \mathscr{L}_{HS}(\mathbb{H})$$

are progressively measurable and satisfy the following assumptions for some constant $\alpha > 0$, adapted $\phi \in L^1_{loc}([0,\infty) \to L^1(\mathbb{P}); \mathrm{d}t)$, $K \in C([0,\infty))$, and strictly positive $\psi \in C([0,\infty))$:

(A2.1) Hemicontinuity. For every $t \geq 0$ and $v_1, v_2, v \in \mathbb{V}$,

$$\mathbb{R} \ni s \mapsto_{\mathbb{V}^*} \langle b(t, v_1 + s v_2), v\rangle_{\mathbb{V}}$$

is continuous.

(A2.2) Monotonicity. For every $v_1, v_2 \in \mathbb{V}, t \geq 0$,

$$2_{\mathbb{V}^*}\langle b(t,v_1) - b(t,v_2), v_1 - v_2\rangle_{\mathbb{V}} + \|\sigma(t,v_1) - \sigma(t,v_2)\|_{HS}^2 \leq K(t)|v_1 - v_2|^2.$$

F.-Y. Wang, *Harnack Inequalities for Stochastic Partial Differential Equations*, SpringerBriefs in Mathematics, DOI 10.1007/978-1-4614-7934-5_2,

(A2.3) Coercivity. For every $t \geq 0, v \in \mathbb{V}$,

$$2_{\mathbb{V}^*}\langle b(t,v),v\rangle_{\mathbb{V}} + \|\sigma(t,v)\|_{HS}^2 \leq \phi(t) + K(t)|v|^2 - \psi(t)\|v\|_{\mathbb{V}}^{\alpha+1}.$$

(A2.4) Growth. For every $u,v \in \mathbb{V}, t \geq 0$,

$$|_{\mathbb{V}^*}\langle b(t,v),u\rangle_{\mathbb{V}}| \leq \phi(t) + K(t)\{\|v\|_{\mathbb{V}}^{\alpha} + \|u\|_{\mathbb{V}}^{\alpha+1} + |u|^2 + |v|^2\}.$$

Definition 2.1.1 *A continuous $\mathbb{H}$-valued adapted process X is called a (strong or variational) solution to (2.1) if*

$$\mathbb{E}\int_0^T (\|X(t)\|_{\mathbb{V}}^{\alpha+1} + |X(t)|^2)\mathrm{d}t < \infty, \ \ T > 0,$$

and $\mathbb{P}$-a.s.

$$X(t) = X(0) + \int_0^t b(s,X(s))\mathrm{d}s + \int_0^t \sigma(s,X(s))\mathrm{d}W(s), \ \ t \geq 0.$$

According to [24, Theorems II.2.1, II.2.2], for every $\mathscr{F}_0$-measurable $X(0) \in L^2(\Omega \to \mathbb{H}; \mathbb{P})$, (2.1) has a unique solution $X = (X(t))_{t\geq 0}$; see also [40, Theorem 2.1] for

$$\mathbf{K} := L^{1+\alpha}([0,T]\times\Omega \to \mathbb{V}; \mathrm{d}t\times\mathbb{P}) \cap L^2([0,T]\times\Omega \to \mathbb{H}; \mathrm{d}t\times\mathbb{P}).$$

Moreover, $|X|^2$ satisfies the Itô formula

$$\mathrm{d}|X(t)|^2 = 2_{\mathbb{V}^*}\langle b(t,X(t)),X(t)\rangle_{\mathbb{V}}\mathrm{d}t + \|\sigma(t,X(t))\|_{HS}^2\mathrm{d}t + 2\langle X(t), \sigma(t,X(t))\mathrm{d}W(t)\rangle,$$

and hence

$$\mathbb{E}\sup_{t\in[0,T]}|X(t)|^2 + \mathbb{E}\int_0^T \|X(t)\|_{\mathbb{V}}^{\alpha+1}\mathrm{d}t < \infty, \ \ T > 0.$$

The study of (2.1) with the above assumptions goes back to [37, 38], and such equations are called (nonlinear) monotone SPDEs. A typical example is that of the following stochastic generalized porous media and fast-diffusion equations. Generalizations with local conditions or for nonmonotone equations can be found in [29, 30, 42].

Example 2.1.1 (Stochastic generalized porous media/fast-diffusion equations) *Let $(\Delta, \mathscr{D}(\Delta))$ be the Dirichlet Laplacian on a bounded open domain $D \subset \mathbb{R}^d$. Let $\alpha > 0$ be a constant. Consider the following PDE (partial differential equation):*

$$\partial_t u = \Delta u^{\alpha}, \ \ t \geq 0, u(t,\cdot)|_{\partial D} = 0,$$

where $u^{\alpha} := |u|^{\alpha}\mathrm{sign}(u)$. This equation is called the (Dirichlet) heat equation if $\alpha = 1$, the fast-diffusion equation if $\alpha \in (0,1)$, and the porous medium equation if $\alpha > 1$. Since $\Delta \leq -\lambda_1$ for some $\lambda_1 > 0$, $\mathscr{D}(\sqrt{-\Delta})$ under the inner product

$$\langle u,v\rangle_{\mathscr{D}} := \int_D \big(\sqrt{-\Delta}u\big)\big(\sqrt{-\Delta}v\big)\mathrm{d}\mathbf{m}$$

is a separable Hilbert space, known as the (first-order) Sobolev space associated to Δ*, where* $\mathbf{m}$ *is the normalized volume measure on D. Let* $\mathbb{H}$ *be the dual space of* $\mathscr{D}(\sqrt{-\Delta})$ *with respect to* $L^2(\mathbf{m})$*, and let* $(W(t))_{\geq 0}$ *be a cylindrical Brownian motion on* $\mathbb{H}$*. Consider the SPDE*

$$\mathrm{d}X(t) = \Delta X(t)^{\alpha}\mathrm{d}t + \sigma(t)\mathrm{d}W(t),$$

where $\sigma : [0,\infty) \to \mathscr{L}_{HS}(\mathbb{H})$ *is measurable such that* $\|\sigma(\cdot)\|_{HS}^2 \in L^1_{loc}([0,\infty);\mathrm{d}t)$*. Let* $\mathbb{V} = \mathbb{H} \cap L^{\alpha+1}(\mathbf{m})$ *be equipped with* $\|\cdot\|_{\mathbb{V}} := |\cdot| + \|\cdot\|_{L^{\alpha+1}(\mathbf{m})}$*. By Sobolev's inequality, if* $\alpha \geq \frac{d-2}{d+2}$*, then* $\mathbb{V} = L^{\alpha+1}(\mathbf{m})$*. It is easy to see that* **(A2.1)–(A2.4)** *hold for* $K = \psi = 1$ *and* $\phi = \|\sigma(\cdot)\|_{HS}^2$*. Therefore, for every* $X(0) \in L^2(\Omega \to \mathbb{H};\mathbb{P})$*, this equation has a unique solution.*

More generally, let $(E,\mathscr{B},\mathbf{m})$ be a probability measure space, and $(L,\mathscr{D}(L))$ a self-adjoint operator on $L^2(\mathbf{m})$ with $L \leq -\lambda_1$ for some constant $\lambda_1 > 0$. Let $\mathbb{H}$ be the dual space of $\mathscr{D}(\sqrt{-L})$ with respect to $L^2(\mathbf{m})$. Let $\mathbb{V} = L^{\alpha+1}(\mathbf{m}) \cap \mathbb{H}$, and let $W(t)$ be a cylindrical Brownian motion on $\mathbb{H}$. Consider the following SPDE on $\mathbb{H}$:

$$\mathrm{d}X(t) = \big\{L\Psi(t,X(t)) + \Phi(t,X(t))\big\}\mathrm{d}t + \sigma(t)\mathrm{d}W(t),$$

where $\sigma : [0,\infty) \to \mathscr{L}_{HS}(\mathbb{H})$ is measurable with $\|\sigma(\cdot)\|_{HS}^2 \in L^1_{loc}([0,\infty);\mathrm{d}t)$, and $\Psi,\Phi : [0,\infty)\times\mathbb{R} \to \mathbb{R}$ are measurable and continuous in the second variable, satisfying the condition that for some functions $\delta,\zeta,\gamma,h \in C([0,\infty))$ with $\delta > 0$,

$$|\Psi(t,s)| + |\Phi(t,s) - h(t)s| \leq \zeta(t)(1+|s|^{\alpha}), \quad s\in\mathbb{R}, t\geq 0, \tag{2.2}$$

$$\begin{aligned} &2\langle \Psi(t,x)-\Psi(t,y), y-x\rangle_2 + 2\langle \Phi(t,x)-\Phi(t,y), (-L)^{-1}(x-y)\rangle_2 \\ &\quad \leq -\delta(t)\|x-y\|_{L^{\alpha+1}(\mathbf{m})}^{\alpha+1} + \gamma(t)|x-y|^2, \quad x,y\in L^{\alpha+1}(\mathbf{m}),\ t\geq 0, \end{aligned} \tag{2.3}$$

where here and in the sequel, $\langle\cdot,\cdot\rangle_2$ denotes the inner product in $L^2(\mathbf{m})$. A very simple example of Ψ,Φ satisfying these conditions is $\Psi(t,s) := |s|^{\alpha-1}s$ and $\Phi(t,s) = \gamma(t)s$. If $\Phi(t,s)$ is nonlinear in s, we also assume that L^{-1} is bounded in $L^{\alpha+1}(\mathbf{m})$, which is the case if L is a Dirichlet operator. Under these conditions, **(A2.1)–(A2.4)** hold for $b(t,v) := L\Psi(t,v) + \Phi(t,v)$ and some ϕ,ψ,K. Thus, for every $X(0) \in L^2(\Omega\to\mathbb{H};\mathbb{P})$, the equation has a unique solution. A simple example for this condition to hold is that $\Psi(\cdot,s) = \eta(\cdot)s^{\alpha}$ for some strictly positive $\eta \in C([0,\infty))$. See Sect. 3 in [40, 55] for more general results where $L^{\alpha+1}(\mathbf{m})$ is replaced by an Orlicz space.

The next example can be found in [27, 39].

Example 2.1.2 (Stochastic *p*-Laplace equation) *Let* $D \subset \mathbb{R}^d$ *be an open domain, let* $\mathbf{m}$ *be the normalized volume measure on D, and let* $p \geq 2$ *be a constant. Let* $\mathbb{H}_0^{1,p}(D)$ *be the closure of* $C_0^{\infty}(D)$ *with respect to the norm*

$$\|f\|_{1,p} := \|f\|_{L^p(\mathbf{m})} + \|\nabla f\|_{L^p(\mathbf{m})}.$$

Let $\mathbb{H} = L^2(\mathbf{m})$ and $\mathbb{V} = \mathbb{H}_0^{1,p}(D)$. By the L^p-Poincaré inequality, there exists a constant $C > 0$ such that $\|f\|_{1,p} \le C\|\nabla f\|_{L^p(\mathbf{m})}$. Consider the SPDE

$$\mathrm{d}X(t) = \mathrm{div}\big(|\nabla X(t)|^{p-2}\nabla X(t)\big)\mathrm{d}t + \sigma(t)\mathrm{d}W(t),$$

where $W(t)$ is a cylindrical Brownian motion on $\mathbb{H}$, and $\sigma : [0,\infty) \to \mathscr{L}_{HS}(\mathbb{H})$ is measurable with $\|\sigma(\cdot)\|_{HS}^2 \in L^1_{loc}([0,\infty);\mathrm{d}t)$. Then **(A2.1)–(A2.4)** *hold for $b(t,v) := \mathrm{div}\big(|\nabla v|^{p-2}\nabla v\big)$ and some ϕ, ψ, K. Thus, for every $\mathscr{F}_0$-measurable $X(0) \in L^2(\Omega \to \mathbb{H};\mathbb{P})$, the equation has a unique solution.*

2.2 Harnack Inequalities for $\alpha \ge 1$

In this section, we consider monotone stochastic equations with deterministic coefficients and additive noise, i.e., equations of the form

$$\mathrm{d}X(t) = b(t,X(t))\mathrm{d}t + \sigma(t)\mathrm{d}W(t), \tag{2.4}$$

where $W(t)$ is a cylindrical Brownian motion on $\mathbb{H}$, $\sigma : [0,\infty) \to \mathscr{L}_{HS}(\mathbb{H})$ is measurable with $\|\sigma\|_{HS} \in L^2_{loc}([0,\infty);\mathrm{d}t)$, and $b : [0,\infty) \times \mathbb{V} \to \mathbb{V}^*$ is measurable such that **(A2.1)–(A2.4)** hold for $\sigma(t,v) := \sigma(t)$ and some $\alpha \ge 1$. For every $x \in \mathbb{H}$, let $X^x(t)$ be the solution with $X(0) = x$. We aim to establish Harnack inequalities for the associated Markov operators $P_t, t > 0$:

$$P_t f(x) := \mathbb{E} f(X^x(t)), \quad f \in \mathscr{B}_b(\mathbb{H}), x \in \mathbb{H},$$

using Theorem 1.1.1. In this case, the idea for the construction of coupling is due to [54], where stochastic generalized porous media equations were considered, and it was extended in [27, 28, 31] to more general equations. The construction of Harnack inequalities for nonlinear monotone SPDEs with multiplicative noise is still open at the moment. Although coupling by change of measure has been constructed in [57] to establish Harnack inequalities for SDEs with multiplicative noise (see Sect. 3.4 below), it is very hard to apply the construction to nonlinear SPDEs with multiplicative noise.

Throughout this section, we assume that the noise is nondegenerate, i.e., $\sigma(t)v=0$ for some $v \in \mathbb{H}$ implies that $v = 0$. For every $v \in \mathbb{H}$, we define the intrinsic norm induced by $\sigma(t)$ by

$$\|v\|_{\sigma(t)} = |\sigma(t)^{-1}v|,$$

where $|\sigma(t)^{-1}v| := |y|$ if $y \in \mathbb{H}$ such that $\sigma(t)y = v$; $|\sigma(t)^{-1}v| = \infty$ if $v \notin \sigma(t)\mathbb{H}$.

The main result of this section is the following.

Theorem 2.2.1 *If there exist a constant $\theta \in [2,\infty) \cap (\alpha - 1,\infty)$ and $\eta, \gamma \in C([0,\infty))$ with $\eta > 0$ such that*

$$2_{\mathbb{V}^*}\langle b(t,u) - b(t,v), u - v\rangle_{\mathbb{V}} \le -\eta(t)\|u - v\|_{\sigma(t)}^{\theta}|u - v|^{\alpha+1-\theta} + \gamma(t)|u - v|^2 \tag{2.5}$$

holds for $t \geq 0$ and $u,v \in \mathbb{V}$, then for every $T > 0, x,y \in \mathbb{H}$, and strictly positive $f \in \mathscr{B}_b(\mathbb{H})$,

$$(P_T f(y))^p \leq (P_T f^p(x)) \exp\left[\frac{p\left(\frac{\theta+2}{\theta+1-\alpha}\right)^{\frac{2(\theta+1)}{\theta}} |x-y|^{\frac{2(\theta+1-\alpha)}{\theta}}}{2(p-1)\left(\int_0^T \eta(t)^{\frac{2}{\theta+2}} e^{-\frac{\theta+1-\alpha}{\theta+2}\int_0^t \gamma(s)ds} dt\right)^{\frac{\theta+2}{\theta}}}\right],$$

$$P_T \log f(y) \leq \log P_T f(x) + \frac{\left(\frac{\theta+2}{\theta+1-\alpha}\right)^{\frac{2(\theta+1)}{\theta}} |x-y|^{\frac{2(\theta+1-\alpha)}{\theta}}}{2\left(\int_0^T \eta(t)^{\frac{2}{\theta+2}} \exp\left[-\frac{\theta+1-\alpha}{\theta+2}\int_0^t \gamma(s)ds\right]dt\right)^{\frac{\theta+2}{\theta}}}.$$

We now explain the idea of the proof using coupling by change of measure. Let $T > 0$ and $x,y \in \mathbb{H}$ be fixed, and let $X(t) = X^x(t)$ be the solution to (2.4) with $X(0) = x$. We intend to construct $Y(t)$ with $Y(0) = y$ such that $Y(T) = X(T)$ and the law of $Y(T)$ under a weighted probability $d\mathbb{Q} := R d\mathbb{P}$ coincides with that of $X^y(T)$. To this end, let $Y(t)$ solve the equation

$$dY(t) = \left\{b(t,Y(t)) + \frac{\xi(t)(X(t)-Y(t))}{|X(t)-Y(t)|^\varepsilon}\right\}dt + \sigma(t)dW(t), \quad Y(0) = y, \tag{2.6}$$

where $\varepsilon \in (0,1)$ and $\xi \in C([0,\infty))$ are to be determined such that $Y(T) = X(T)$, and if $u = v$, we set $\frac{u-v}{|u-v|^\varepsilon} = 0$. As shown in the proof of [54, Theorem A.2], we have that **(A2.1)–(A2.4)** hold with $\sigma(t,v) = \sigma(t)$ and $b(t,v)$ replaced by

$$\tilde{b}(t,v) := b(t,v) + \frac{\xi(t)(X(t)-v)}{|X(t)-v|^\varepsilon}, \quad t \geq 0, v \in \mathbb{V}.$$

Therefore, (2.6) has a unique solution.

Now we aim to choose ξ and ε such that

$$\tau := \inf\{t \geq 0 : X(t) = Y(t)\} \leq T,$$

which implies that $X(T) = Y(T)$ by the uniqueness of solutions to (2.4) for $t \geq \tau$.

Lemma 2.2.2 *Let $\xi \in C([0,\infty))$ and $\varepsilon \in (0,1)$ be such that*

$$\int_0^T \xi(t) e^{-\frac{\varepsilon}{2}\int_0^t \gamma(s)ds} dt \geq \frac{1}{\varepsilon}|x-y|^\varepsilon. \tag{2.7}$$

Then $X(T) = Y(T)$.

Proof. By (2.5) and Itô's formula, we have

$$d|X(t)-Y(t)|^2 \leq \left(\gamma(t)|X(t)-Y(t)|^2 - 2\xi(t)|X(t)-Y(t)|^{2-\varepsilon}\right)dt, \quad t < \tau.$$

Then

$$\begin{aligned}
&\mathrm{d}\Big\{|X(t)-Y(t)|^2\mathrm{e}^{-\int_0^t \gamma(s)\mathrm{d}s}\Big\}^{\frac{\varepsilon}{2}}\\
&=\frac{\varepsilon}{2}\Big\{|X(t)-Y(t)|^2\mathrm{e}^{-\int_0^t \gamma(s)\mathrm{d}s}\Big\}^{\frac{\varepsilon-2}{2}}\mathrm{d}\Big\{|X(t)-Y(t)|^2\mathrm{e}^{-\int_0^t \gamma(s)\mathrm{d}s}\Big\}\\
&\le -\varepsilon\xi(t)\mathrm{e}^{-\frac{\varepsilon}{2}\int_0^t \gamma(s)\mathrm{d}s}\mathrm{d}t,\ \ t<\tau.
\end{aligned}$$

This implies

$$|X(t)-Y(t)|^{\varepsilon}\mathrm{e}^{-\frac{\varepsilon}{2}\int_0^t \gamma(s)\mathrm{d}s}\le |x-y|^{\varepsilon}-\varepsilon\int_0^t \xi(s)\mathrm{e}^{-\frac{\varepsilon}{2}\int_0^s \gamma(r)\mathrm{d}r}\mathrm{d}s,\ \ t<\tau. \tag{2.8}$$

If $\tau>T$, then $|X(T)-Y(T)|>0$, but by (2.7) and (2.8), we have $|X(T)-Y(T)|^{\varepsilon}\le 0$, which is a contradiction. $\square$

Proof of Theorem 2.2.1.

(a) Take

$$\begin{aligned}
&\varepsilon=\frac{\theta+1-\alpha}{\theta+2},\\
&\xi(t)=\frac{(\theta+2)\eta(t)^{\frac{2}{\theta+2}}\exp\big[-\frac{\theta+1-\alpha}{2(\theta+2)}\int_0^t\gamma(s)\mathrm{d}s\big]}{(\theta+1-\alpha)\int_0^T\eta(t)^{\frac{2}{\theta+2}}\exp\big[-\frac{\theta+1-\alpha}{\theta+2}\int_0^t\gamma(s)\mathrm{d}s\big]\mathrm{d}t}|x-y|^{\frac{\theta+1-\alpha}{\theta+2}},\ \ t\ge 0.
\end{aligned}$$

Then

$$\int_0^T\xi(t)\mathrm{e}^{-\frac{\varepsilon}{2}\int_0^t\gamma(s)\mathrm{d}s}\mathrm{d}t=\frac{\theta+2}{\theta+1-\alpha}|x-y|^{\frac{\theta+1-\alpha}{\theta+2}}=\frac{|x-y|^{\varepsilon}}{\varepsilon}.$$

By Lemma 2.2.2, we have $X(T)=Y(T)$.

(b) To formulate the changed probability $\mathrm{d}\mathbb{Q}=R\mathrm{d}\mathbb{P}$, we rewrite (2.6) by

$$\mathrm{d}Y(t)=b(t,Y(t))\mathrm{d}t+\sigma(t)\mathrm{d}\tilde{W}(t),\ \ Y(0)=y,$$

where

$$\tilde{W}(t):=W(t)+\int_0^{\tau\wedge t}\frac{\xi(s)\sigma(s)^{-1}(X(s)-Y(s))}{|X(s)-Y(s)|^{\varepsilon}}\mathrm{d}s.$$

We shall prove

$$\begin{aligned}
&\int_0^{\tau}\Big|\xi(s)\frac{\sigma(s)^{-1}(X(s)-Y(s))}{|X(s)-Y(s)|^{\varepsilon}}\Big|^2\mathrm{d}s\\
&\le\frac{\big(\frac{\theta+2}{\theta+1-\alpha}\big)^{\frac{2(\theta+1)}{\theta}}|x-y|^{\frac{2(\theta+1-\alpha)}{\theta}}}{\big(\int_0^T\eta(t)^{\frac{2}{\theta+2}}\exp\big[-\frac{\theta+1-\alpha}{\theta+2}\int_0^t\gamma(s)\mathrm{d}s\big]\mathrm{d}t\big)^{\frac{\theta+2}{\theta}}},
\end{aligned} \tag{2.9}$$

which ensures by Girsanov's theorem that $\tilde{W}(t)$ is a cylindrical Brownian motion on $\mathbb{H}$ under $\mathrm{d}\mathbb{Q}:=R\mathrm{d}\mathbb{P}$, where

$$R := \exp\Big[-\int_0^\tau \frac{\xi(s)}{|X(s)-Y(s)|^\varepsilon}\langle \sigma(s)^{-1}(X(s)-Y(s)), \mathrm{d}W(s)\rangle$$
$$-\frac{1}{2}\int_0^\tau \Big|\xi(s)\frac{\sigma(s)^{-1}(X(s)-Y(s))}{|X(s)-Y(s)|^\varepsilon}\Big|^2 \mathrm{d}s\Big]. \tag{2.10}$$

By (2.5) and Itô's formula, we have

$$\mathrm{d}|X(t)-Y(t)|^2 \leq \Big\{-\eta(t)\|X(t)-Y(t)\|_{\sigma(t)}^\theta |X(t)-Y(t)|^{\alpha+1-\theta} + \gamma(t)|X(t)-Y(t)|^2\Big\}\mathrm{d}t$$

for $t<\tau$. Then,

$$\begin{aligned}
&\mathrm{d}\Big\{|X(t)-Y(t)|^2 \mathrm{e}^{-\int_0^t \gamma(s)\mathrm{d}s}\Big\}^\varepsilon \\
&\leq -\varepsilon\Big\{|X(t)-Y(t)|^2\mathrm{e}^{-\int_0^t\gamma(s)\mathrm{d}s}\Big\}^{\varepsilon-1}\eta(t)\|X(t)-Y(t)\|_{\sigma(t)}^\theta \\
&\quad\times |X(t)-Y(t)|^{\alpha+1-\theta}\mathrm{e}^{-\int_0^t\gamma(s)\mathrm{d}s}\mathrm{d}t \\
&= -\varepsilon\eta(t)\mathrm{e}^{-\varepsilon\int_0^t\gamma(s)\mathrm{d}s}\frac{\|X(t)-Y(t)\|_{\sigma(t)}^\theta}{|X(t)-Y(t)|^{\theta\varepsilon}}\mathrm{d}t,\ \ t<\tau,
\end{aligned}$$

where in the last step we have used $\varepsilon = \frac{\theta+1-\alpha}{\theta+2}$. Thus, again using $\varepsilon = \frac{\theta+1-\alpha}{\theta+2}$, we obtain

$$\int_0^\tau \eta(t)\mathrm{e}^{-\varepsilon\int_0^t\gamma(s)\mathrm{d}s}\frac{\|X(t)-Y(t)\|_{\sigma(t)}^\theta}{|X(t)-Y(t)|^{\theta\varepsilon}}\mathrm{d}t \leq \frac{\theta+2}{\theta+1-\alpha}|x-y|^{\frac{2(\theta+1-\alpha)}{\theta+2}}.$$

Since $\theta \geq 2$ and $\tau \leq T$, by Hölder's inequality, we obtain from this that

$$\begin{aligned}
&\int_0^\tau \Big|\xi(s)\frac{\sigma(s)^{-1}(X(s)-Y(s))}{|X(s)-Y(s)|^\varepsilon}\Big|^2\mathrm{d}s = \int_0^\tau \xi(s)^2\frac{\|X(s)-Y(s)\|_{\sigma(s)}^2}{|X(s)-Y(s)|^{2\varepsilon}}\mathrm{d}s \\
&\leq \Big(\int_0^\tau \eta(t)\mathrm{e}^{-\varepsilon\int_0^t\gamma(s)\mathrm{d}s}\frac{\|X(t)-Y(t)\|_{\sigma(t)}^\theta}{|X(t)-Y(t)|^{\theta\varepsilon}}\mathrm{d}t\Big)^{\frac{2}{\theta}}\Big(\int_0^T \mathrm{e}^{\frac{2\varepsilon}{\theta-2}\int_0^t\gamma(s)\mathrm{d}s}\frac{\xi(t)^{\frac{2\theta}{\theta-2}}}{\eta(t)^{\frac{2}{\theta-2}}}\mathrm{d}t\Big)^{\frac{\theta-2}{\theta}} \\
&\leq \Big(\frac{2+\theta}{\theta+1-\alpha}\Big)^{\frac{2}{\theta}}|x-y|^{\frac{4(\theta+1-\alpha)}{\theta(\theta+2)}}\Big(\int_0^T \mathrm{e}^{\frac{2(\theta+1-\alpha)}{\theta^2-4}\int_0^t\gamma(s)\mathrm{d}s}\frac{\xi(t)^{\frac{2\theta}{\theta-2}}}{\eta(t)^{\frac{2}{\theta-2}}}\mathrm{d}t\Big)^{\frac{\theta-2}{\theta}}.
\end{aligned} \tag{2.11}$$

By the definition of $\xi(t)$, we have

$$\begin{aligned}
&\int_0^T \mathrm{e}^{\frac{2(\theta+1-\alpha)}{\theta^2-4}\int_0^t\gamma(s)\mathrm{d}s}\frac{\xi(t)^{\frac{2\theta}{\theta-2}}}{\eta(t)^{\frac{2}{\theta-2}}}\mathrm{d}t \\
&= \frac{\Big(\frac{\theta+2}{\theta+1-\alpha}|x-y|^{\frac{\theta+1-\alpha}{\theta+2}}\Big)^{\frac{2\theta}{\theta-2}}}{\Big(\int_0^T\eta(t)^{\frac{2}{\theta+2}}\exp\big[-\frac{\theta+1-\alpha}{\theta+2}\int_0^t\gamma(s)\mathrm{d}s\big]\mathrm{d}t\Big)^{\frac{\theta+2}{\theta-2}}}.
\end{aligned}$$

Combining this with (2.11), we prove (2.9).

(c) By Theorem 1.1.1, the proof is finished, since according to (2.10) and (2.9), we have

$$\Big(\mathbb{E}R^{\frac{p}{p-1}}\Big)^{p-1} \le \operatorname*{ess\,sup}_{\Omega} \exp\left[\frac{p}{2(p-1)}\int_0^\tau \Big|\xi(s)\frac{\sigma(s)^{-1}(X(s)-Y(s))}{|X(s)-Y(s)|^{\varepsilon}}\Big|^2 \mathrm{d}s\right]$$

$$\le \exp\left[\frac{p\big(\frac{\theta+2}{\theta+1-\alpha}\big)^{\frac{2(\theta+1)}{\theta}}|x-y|^{\frac{2(\theta+1-\alpha)}{\theta}}}{2(p-1)\big(\int_0^T \eta(t)^{\frac{2}{\theta+2}}\exp\big[-\frac{\theta+1-\alpha}{\theta+2}\int_0^t \gamma(s)\mathrm{d}s\big]\mathrm{d}t\big)^{\frac{\theta+2}{\theta}}}\right]$$

and

$$\mathbb{R}[R\log R] = \frac{1}{2}\mathbb{E}_{\mathbb{Q}}\int_0^\tau \Big|\xi(s)\frac{\sigma(s)^{-1}(X(s)-Y(s))}{|X(s)-Y(s)|^{\varepsilon}}\Big|^2 \mathrm{d}s$$

$$\le \frac{\big(\frac{\theta+2}{\theta+1-\alpha}\big)^{\frac{2(\theta+1)}{\theta}}|x-y|^{\frac{2(\theta+1-\alpha)}{\theta}}}{2\big(\int_0^T \eta(t)^{\frac{2}{\theta+2}}\exp\big[-\frac{\theta+1-\alpha}{\theta+2}\int_0^t \gamma(s)\mathrm{d}s\big]\mathrm{d}t\big)^{\frac{\theta+2}{\theta}}}.$$

□

As applications of the Harnack inequalities, we consider estimates of P_t in terms of its invariant probability measure. To this end, we assume that the equation is time-homogeneous, i.e., $b(t,\cdot)$ and $\sigma(t)$ are independent of t. In this case, P_t is a Markov semigroup.

Proposition 2.2.3 *Assume that $b(t,\cdot)$ and $\sigma(t)$ are independent of t such that* **(A2.1)–(A2.4)** *hold with constants ϕ, K, Ψ such that $\Psi > 0$. If the embedding $\mathbb{V}\subset\mathbb{H}$ is compact and*

$$K|v|^2 - \Psi\|v\|_{\mathbb{V}}^{\alpha+1} \le C - \delta\|v\|_{\mathbb{V}}^{\alpha+1},\quad v\in\mathbb{V} \tag{2.12}$$

holds for some constants $C,\delta>0$, then P_t has an invariant probability measure μ such that

$$\int_{\mathbb{H}}\Big(\|\cdot\|_{\mathbb{V}}^{\alpha+1} + \mathrm{e}^{c|\cdot|^{\alpha+1}}\Big)\mathrm{d}\mu < \infty \tag{2.13}$$

holds for some constant $c>0$, where $\|v\|_{\mathbb{V}} := \infty$ for $v\in\mathbb{H}\setminus\mathbb{V}$.

Proof. By **(A2.3)**, (2.12), and Itô's formula, we obtain

$$\mathrm{d}|X(t)|^2 \le \{C_1 - \delta\|X(t)\|_{\mathbb{V}}^{\alpha+1}\}\mathrm{d}t + 2\langle X(t), \sigma\mathrm{d}W(t)\rangle \tag{2.14}$$

for some constants $C_1,\delta>0$. This implies that

$$\frac{1}{T}\mathbb{E}\int_0^T \|X^0(t)\|_{\mathbb{V}}^{\alpha+1}\mathrm{d}t \le \frac{C_1}{\delta},\quad T>0.$$

Since the embedding $\mathbb{V}\subset\mathbb{H}$ is compact, the sequence $\{\frac{1}{n}\int_0^n \delta_0 P_t \mathrm{d}t\}_{n\ge 1}$ is tight, where δ_0 is the Dirac measure at the point $0\in\mathbb{H}$. By a standard argument, a weak limit μ of a subsequence is an invariant probability measure and $\mu(\|\cdot\|_{\mathbb{V}}^{\alpha+1}) \leq \frac{C_1}{\delta}$. It remains to find a constant $c > 0$ such that $\mu(e^{c|\cdot|^{\alpha+1}}) < \infty$. For every $\varepsilon > 0$, it follows from (2.14) that

$$\begin{aligned} de^{\varepsilon|X(t)|^{\alpha+1}} \leq\ & dM_t + \frac{(\alpha+1)\varepsilon}{2} e^{\varepsilon|X(t)|^{\alpha+1}} |X(t)|^{\alpha-1} \\ & \times \big(C_2 + (\alpha+1)\varepsilon|X(t)|^{\alpha+1} - \delta\|X(t)\|_{\mathbb{V}}^{\alpha+1}\big)\,dt \end{aligned}$$

holds for some local martingale M_t and some constant $C_2 > 0$. Since $|\cdot| \leq c_0\|\cdot\|_{\mathbb{V}}$ holds for some constant $c_0 > 0$, we see that for small enough $\varepsilon > 0$,

$$\begin{aligned} & e^{\varepsilon|X(t)|^{\alpha+1}} |X(t)|^{\alpha-1} \big(C_2 + (\alpha+1)\varepsilon|X(t)|^{\alpha+1} - \delta\|X(t)\|_{\mathbb{V}}^{\alpha+1}\big) \\ & \leq \tilde{C} - \tilde{\delta}|X(t)|^{2\alpha} e^{\varepsilon|X(t)|^{\alpha+1}} \leq C' - \delta' e^{\varepsilon|X(t)|^{\alpha+1}} \end{aligned}$$

holds for some constants $\tilde{C}, \tilde{\delta}, C', \delta' > 0$. Therefore, for small $\varepsilon > 0$,

$$de^{\varepsilon|X(t)|^{\alpha+1}} - dM_t \leq \big\{\tilde{C} - \tilde{\delta}|X(t)|^{2\alpha} e^{\varepsilon|X(t)|^{\alpha+1}}\big\}dt \leq \big\{C' - \delta' e^{\varepsilon|X(t)|^{\alpha+1}}\big\}dt. \quad (2.15)$$

In particular, there exists a constant $C'' > 0$ such that for small enough $\varepsilon > 0$, we have

$$\mathbb{E}e^{\varepsilon|X^0(t)|^{\alpha+1}} \leq C'', \ t \geq 0.$$

This implies that $\mu(e^{\varepsilon|\cdot|^{\alpha+1}}) < \infty$. □

Corollary 2.2.4 *In the situation of Proposition* 2.2.3 *and under the assumption that* (2.5) *holds for some constant functions* $\eta > 0$ *and* γ, *we have that* P_t *is strong Feller and has a unique invariant probability measure* μ, *and* μ *has full support on* $\mathbb{H}$. *Moreover:*

(1) *If* $\alpha = 1$ *and* $\gamma \leq 0$, *then* P_t *is hyperbounded, i.e.,* $\|P_t\|_{L^2(\mu)\to L^4(\mu)} < \infty$ *for some* $t > 0$.
(2) *If* $\alpha > 1$, *then* P_t *is ultracontractive with*

$$\|P_t\|_{L^2(\mu)\to L^\infty(\mu)} \leq \exp\Big[C + Ct^{-\frac{\alpha+1}{\alpha-1}}\Big], \ t > 0,$$

holding for some constant $C > 0$.
(3) *Let* P_t^* *be the adjoint operator of* P_t *in* $L^2(\mu)$. *Let* W_1^ρ *denote the* L^1-*transportation cost induced by the cost function* $\rho(x,y) := |x-y|^{\frac{2(\theta+1-\alpha)}{\theta}}$; *see Theorem* 1.4.2. *Then*

$$\mu((P_T^*f)\log P_T^*f) \leq \frac{\theta+2}{2\eta^{\frac{2}{\theta}}(\theta+1-\alpha)} \left(\frac{\gamma}{1-\exp\big[-\frac{\gamma(\theta+1-\alpha)}{\theta+2}T\big]}\right)^{\frac{\theta+2}{\theta}} W_1^\rho(f\mu,\mu)$$

holds for all $T > 0$ *and* $f \geq 0$ *with* $\mu(f) = 1$.

Proof. (a) By Theorem 2.2.1 with constants η, γ, and $p=2$, (1.2) holds for $P=P_T$, $\Phi(r)=r^2$, and

$$\Psi(x,y)=c_1\left(\frac{\gamma}{1-e^{-c_2\gamma T}}\right)^{\frac{\theta+2}{\theta}}|x-y|^{\frac{2(\theta+1-\alpha)}{\theta}}$$

for some constants $c_1,c_2>0$, i.e.,

$$(P_Tf(x))^2\le(P_Tf^2(y))\exp\left[c_1\left(\frac{\gamma}{1-e^{-c_2\gamma T}}\right)^{\frac{\theta+2}{\theta}}|x-y|^{\frac{2(\theta+1-\alpha)}{\theta}}\right] \tag{2.16}$$

holds for $T>0$, $x,y\in\mathbb{H}$, and $f\in\mathscr{B}_b^+(\mathbb{H})$. Then by Theorem 1.4.1(3) and Proposition 2.2.3, P_t has a unique invariant measure μ. To see that μ has full support on $\mathbb{H}$, it suffices to show that $\mu(B(x,\varepsilon))>0$ for every $x\in\mathbb{H}$ and $\varepsilon>0$. Applying (2.16) to $f=1_{B(x,\varepsilon)}$, we obtain

$$(P_T1_{B(x,\varepsilon)}(x))^2\exp\left[-c_1\left(\frac{\gamma}{1-e^{-c_2\gamma T}}\right)^{\frac{\theta+2}{\theta}}|x-y|^{\frac{2(\theta+1-\alpha)}{\theta}}\right]\le P_T1_{B(x,\varepsilon)}(y).$$

If $\mu(B(x,\varepsilon))=0$, then taking the integral of both sides with respect to $\mu(\mathrm{d}y)$ yields

$$\mathbb{P}(|X^x(T)-x|\ge\varepsilon)=1-P_T1_{B(x,\varepsilon)}(x)=1,\quad T>0.$$

This is impossible, since the solution is continuous, so that $X^x(T)\to x$ as $T\to0$.

(b) Let $\alpha=1$ and $\gamma\le0$. Then (2.16) implies

$$(P_Tf(x))^2\le(P_Tf^2(y))\exp\left[\frac{c|x-y|^2}{T^{\frac{\theta+2}{\theta}}}\right]$$

for some constant $c>0$. So by Theorem 1.4.1(6), we obtain

$$\begin{aligned}\sup_{\mu(f^2)\le1}|P_Tf(x)| &\le\frac{1}{\int_{B(0,|x|+1)}\exp[-\frac{c(1+2|x|)^2}{T^{\frac{\theta+2}{\theta}}}]\mu(\mathrm{d}y)}\\ &\le\frac{1}{\mu(B(0,1))}\exp\left[\frac{c(1+2|x|)^2}{T^{\frac{\theta+2}{\theta}}}\right].\end{aligned}$$

Combining with (2.13), we see that for sufficiently large $T>0$, $\|P_T\|_{L^2(\mu)\to L^4(\mu)}<\infty$.

(c) Let $\alpha>1$. Then (2.16) implies

$$(P_Tf(x))^2\le(P_Tf^2(y))\exp\left[\frac{c|x-y|^{\frac{2(\theta+1-\alpha)}{\theta}}}{(1\wedge T)^{\frac{\theta+2}{\theta}}}\right]$$

for some constant $c>0$. This implies

$$\sup_{\mu(f^2)\le 1} |P_T f(x)| \le C_1 \exp\left[\frac{C_2|x|^{\frac{2(\theta+1-\alpha)}{\theta}}}{(1\wedge T)^{\frac{\theta+2}{\theta}}}\right]$$

for some constants $C_1, C_2 > 0$. So

$$\sup_{\mu(f^2)\le 1} |P_T f(x)| \le \sup_{\mu(f^2)\le 1} \left|P_{\frac{T}{2}}(P_{\frac{T}{2}} f)(x)\right| \le C_1 \mathbb{E}\exp\left[\frac{C_2|X^x(\frac{T}{2})|^{\frac{2(\theta+1-\alpha)}{\theta}}}{(1\wedge T)^{\frac{\theta+2}{\theta}}}\right]. \tag{2.17}$$

By (2.15) and Jensen's inequality, we see that $h(t) := \mathbb{E}e^{\varepsilon|X^x(t)|^{\alpha+1}}$ satisfies

$$h'(t) \le \tilde{C} - \tilde{\delta}\mathbb{E}\left\{|X^x(t)|^{2\alpha} e^{\varepsilon|X^x(t)|^{\alpha+1}}\right\} \le \tilde{C} - \tilde{\delta}\varepsilon^{-\frac{2\alpha}{1+\alpha}} h(t)\{\log h(t)\}^{\frac{2\alpha}{\alpha+1}}.$$

Since $\frac{2\alpha}{\alpha+1} > 1$, this implies

$$\mathbb{E}e^{\varepsilon|X^x(t)|^{\alpha+1}} \le \exp\left[c + ct^{-\frac{\alpha+1}{\alpha-1}}\right], \quad t > 0, x \in \mathbb{H}$$

for some constant $c > 0$. Noting that there exists a constant $c' > 0$ such that

$$\frac{C_2|X^x(t)|^{\frac{2(\theta+1-\alpha)}{\theta}}}{(1\wedge t)^{\frac{\theta+2}{\theta}}} \le \varepsilon|X^x(t)|^{\alpha+1} + \frac{c'}{(1\wedge t)^{\frac{\alpha+1}{\alpha-1}}}$$

holds for some constant $c' > 0$ and all $t > 0$, we arrive at

$$\mathbb{E}\exp\left[\frac{C_2|X^x(\frac{T}{2})|^{\frac{2(\theta+1-\alpha)}{\theta}}}{(1\wedge T)^{\frac{\theta+2}{\theta}}}\right] \le \exp\left[c'' + c''T^{-\frac{\alpha+1}{\alpha-1}}\right]$$

for some constant $c'' > 0$ and all $x \in \mathbb{H}$, $T > 0$. Combining this with (2.17), we prove (2).

(d) The entropy–cost inequality in (3) follows from Theorem 1.4.2(3) and the log-Harnack inequality in Theorem 2.2.1 with constants η and γ. □

2.3 Harnack Inequalities for $\alpha \in (0,1)$

When $\alpha \in (0,1)$, for the typical model that $b(t,u) = \Delta u^\alpha$ as in Example 2.1.1, the upper bound of ${}_{\mathbb{V}^*}\langle b(t,u) - b(t,v), u - v\rangle_{\mathbb{V}}$ behaves like $-\mathbf{m}\big(|u-v|^2(|u|\vee|v|)^{\alpha-1}\big)$, so that the condition (2.5) does not hold. In this section, we investigate Harnack inequalities for (2.4) under assumptions **(A2.1)**–**(A2.4)** with $\alpha \in (0,1)$ by introducing the following conditions (2.18) and (2.19) to replace (2.5): for some measurable function $h : \mathbb{V} \to (0,\infty)$, some constant $\theta \ge \frac{4}{\alpha+1}$, some $\gamma \in C([0,\infty))$, and strictly positive $q, \delta, \eta \in C([0,\infty))$,

$$2_{\mathbb{V}^*}\langle b(t,u),u\rangle_{\mathbb{V}} + \|\sigma(t)\|_{HS}^2 \le q(t) - \delta(t)h(u)^{\alpha+1} + \gamma(t)|u|^2, \tag{2.18}$$

$$\begin{aligned} &2_{\mathbb{V}^*}\langle b(t,u)-b(t,v),u-v\rangle_{\mathbb{V}} \\ &\le -\frac{\eta(t)\|u-v\|_{\sigma(t)}^{\theta}}{|u-v|^{\theta-2}(h(u)\vee h(v))^{1-\alpha}} + \gamma(t)|u-v|^2 \end{aligned} \tag{2.19}$$

hold for $t \ge 0$ and $u,v \in \mathbb{V}$.

For any constants $\theta \ge \frac{4}{\alpha+1}$, $p > 1$, $T > 0$, and $x,y \in \mathbb{H}$, let

$$\tilde{q}(t) = q(t)\mathrm{e}^{-\int_0^t \gamma(s)\mathrm{d}s}, \quad \tilde{\delta}(t) = \delta(t)\mathrm{e}^{-\int_0^t \gamma(s)\mathrm{d}s}, \quad t \in [0,T].$$

$$c_1(\theta,T) = \int_0^T \eta(t)^{\frac{2(\alpha+1)}{4+\theta+\theta\alpha}} \mathrm{e}^{-\frac{\theta(\alpha+1)}{\theta\alpha+\theta+4}\int_0^t \gamma(s)\mathrm{d}s}\mathrm{d}t,$$

$$\begin{aligned} c_2(\theta,T,x,y) = \frac{1}{\min_{[0,T]}\tilde{\delta}} \bigg\{ & |x|^2 + |y|^2 + 2\int_0^T \tilde{q}(t)\mathrm{d}t \\ & + \frac{(\theta+2)^2}{\theta}\left(\frac{2}{\theta+4}\right)^{\frac{\theta+4}{\theta+2}} |x-y|^{\frac{\theta}{\theta+2}} \left(|y|^2 + \int_0^T \tilde{q}(t)\mathrm{d}t\right)^{\frac{\theta+4}{2(\theta+2)}} \bigg\}. \end{aligned}$$

Theorem 2.3.1 *Let $\alpha \in (0,1)$ and assume that* (2.18) *and* (2.19) *hold for some constant $\theta \ge \frac{4}{\alpha+1}$, measurable $h : \mathbb{V} \to (0,\infty)$, $\gamma \in C([0,\infty))$, and strictly positive $q,\delta,\eta \in C([0,\infty))$.*

(1) *For every $T > 0, x,y \in \mathbb{H}$ and strictly positive $f \in \mathscr{B}_b(\mathbb{H})$,*

$$P_T \log f(y) - \log P_T f(x) \le \frac{|x-y|^2}{2}\left(\frac{\theta+2}{\theta}\right)^{\frac{2(\theta+1)}{\theta}} \frac{c_2(\theta,T,x,y)^{\frac{2(1-\alpha)}{\theta(\alpha+1)}}}{c_1(\theta,T)^{\frac{\theta(\alpha+1)+4}{\theta(\alpha+1)}}}.$$

Consequently, there exists a constant $C > 0$ such that

$$P_T \log f(y) - \log P_T f(x) \le \frac{C|x-y|^2(|x|^2+|y|^2+1\wedge T)^{\frac{2(1-\alpha)}{\theta(1+\alpha)}}}{(T\wedge 1)^{\frac{\theta(\alpha+1)+4}{\theta(\alpha+1)}}}.$$

(2) *There exists a constant $C > 0$ such that for every $T > 0$, $p > 1$, $x,y \in \mathbb{H}$, and $f \in \mathscr{B}_b^+(\mathbb{H})$,*

$$\begin{aligned} &(P_T f(y))^p \le (P_T f^p(x)) \\ &\times \exp\left[\frac{Cp}{p-1}\left(\frac{|x-y|^2(1+|x|^2+|y|^2)}{(T\wedge 1)^{\frac{\theta(\alpha+1)+4}{\theta(\alpha+1)}}} + \left(\frac{p}{p-1}\right)^{\frac{4(1-\alpha)}{\alpha(\theta+2)+\theta-2}} \frac{|x-y|^{\frac{2\theta(\alpha+1)}{\alpha(\theta+2)+\theta-2}}}{(1\wedge T)^{\frac{\theta(\alpha+1)+4}{\alpha(\theta+2)+\theta-2}}}\right)\right]. \end{aligned}$$

To prove this theorem using coupling by change of measure, let $X(t) = X^x(t)$ solve (2.4) with $X(0) = x$, and let $Y(t)$ solve (2.6) for some $\varepsilon \in (0,1)$ and

$$\xi(t) := \frac{|x-y|^{\varepsilon}\eta(t)^{\alpha_2}}{\varepsilon c_1(\theta,T)}\mathrm{e}^{-\alpha_1\int_0^t \gamma(s)\mathrm{d}s}, \quad t \ge 0,$$

where

$$\varepsilon = \frac{\theta}{\theta+2}, \quad \alpha_1 = \frac{\theta(\theta\alpha+\theta+4\alpha)}{2(\theta+2)(\theta\alpha+\theta+4)}, \quad \alpha_2 = \frac{2(\alpha+1)}{4+\theta+\theta\alpha}.$$

Then it is easy to see that

$$\int_0^T \xi(t)\mathrm{e}^{-\frac{\varepsilon}{2}\int_0^t \gamma(s)\mathrm{d}s}\mathrm{d}t = \frac{|x-y|^{\varepsilon}}{\varepsilon}, \tag{2.20}$$

so that $X(T) = Y(T)$ (i.e., $\tau \le T$) according to Lemma 2.2.2. Moreover,

$$c_1(\theta,T) = \int_0^T \eta(t)^{\alpha_2}\mathrm{e}^{-(\alpha_1+\frac{\theta}{2(\theta+2)})\int_0^t \gamma(s)\mathrm{d}s}\mathrm{d}t. \tag{2.21}$$

Next, it is easy to see from Itô's formula and (2.18) that

$$\mathbb{E}\int_0^t \{h(X(s)) \vee h(Y(s))\}^{\alpha+1}\mathrm{d}s < \infty, \quad t > 0.$$

Since $\alpha+1 \ge \frac{2(1-\alpha)}{\theta-2}$, we have $S(t) := \int_0^t \{h(X(s)) \vee h(Y(s))\}^{\frac{2(1-\alpha)}{\theta-2}}\mathrm{d}s < \infty$ for all $t > 0$. Let

$$\tau_n = \tau \wedge \inf\{t \ge 0 : S(t) \ge n\}, \quad n \ge 1.$$

We have $\tau_n \uparrow \tau$ as $n \uparrow \infty$.

Lemma 2.3.2 *Assume* (2.18) *and* (2.19). *For every* $n \ge 1$,

$$R_n := \exp\bigg[-\int_0^{\tau_n} \frac{\xi(s)}{|X(s)-Y(s)|^{\varepsilon}}\langle \sigma(s)^{-1}(X(s)-Y(s)), \mathrm{d}W(s)\rangle \\ -\frac{1}{2}\int_0^{\tau_n}\Big|\xi(s)\frac{\sigma(s)^{-1}(X(s)-Y(s))}{|X(s)-Y(s)|^{\varepsilon}}\Big|^2\mathrm{d}s\bigg]$$

is a well-defined probability density such that

$$\mathbb{E}\{R_n \log R_n\} \\ \le \frac{|x-y|^2}{2}\Big(\frac{\theta+2}{\theta}\Big)^{\frac{2(\theta+1)}{\theta}} \frac{\big(\int_0^T \mathbb{E}_{\mathbb{Q}_n}\big(h(X(t\wedge\tau_n)) \vee h(Y(t\wedge\tau_n))\big)^{\alpha+1}\mathrm{d}t\big)^{\frac{2(1-\alpha)}{\theta(\alpha+1)}}}{c_1(\theta,T)^{\frac{\theta(\alpha+1)+4}{\theta(\alpha+1)}}},$$

and for every $p > 1$,

$$(\mathbb{E}R_n^{\frac{p}{p-1}})^{p-1} \\ \le \bigg(\mathbb{E}_{\mathbb{Q}_n}\exp\bigg[\frac{c(\theta,p)|x-y|^2}{c_1(\theta,T)^{\frac{\theta(\alpha+1)+4}{\theta(\alpha+1)}}}\Big(\int_0^{\tau_n}(h(X(t)) \vee h(Y(t)))^{\alpha+1}\mathrm{d}t\Big)^{\frac{2(1-\alpha)}{\theta(\alpha+1)}}\bigg]\bigg)^{\frac{p-1}{2}},$$

where

$$\mathrm{d}\mathbb{Q}_n = R_n \mathrm{d}\mathbb{P}, \quad c(\theta,p) := \frac{p+1}{(p-1)^2}\left(\frac{\theta+2}{\theta}\right)^{\frac{2(\theta+1)}{\theta}}.$$

Proof. By (2.19) and Itô's formula, we have

$$\mathrm{d}|X(t)-Y(t)|^2 \le -\eta(t)\frac{\|X(t)-Y(t)\|_{\sigma(t)}^{\theta}|X(t)-Y(t)|^{2-\theta}}{(h(X(t))\vee h(Y(t)))^{1-\alpha}}\mathrm{d}t + \gamma(t)|X(t)-Y(t)|^2\mathrm{d}t$$

for $t<\tau$. Since $\theta-2+2(1-\varepsilon)=\varepsilon\theta$, this implies

$$\mathrm{d}\Big\{|X(t)-Y(t)|^2\mathrm{e}^{-\int_0^t\gamma(s)\mathrm{d}s}\Big\}^{\varepsilon} \le -\frac{\varepsilon\eta(t)\|X(t)-Y(t)\|_{\sigma(t)}^{\theta}\mathrm{e}^{-\varepsilon\int_0^t\gamma(s)\mathrm{d}s}}{|X(t)-Y(t)|^{\varepsilon\theta}(h(X(t))\vee h(Y(t)))^{1-\alpha}}\mathrm{d}t$$

for $t<\tau$. So,

$$\int_0^{\tau}\frac{\eta(t)\|X(t)-Y(t)\|_{\sigma(t)}^{\theta}\mathrm{e}^{-\varepsilon\int_0^t\gamma(s)\mathrm{d}s}}{|X(t)-Y(t)|^{\varepsilon\theta}(h(X(t))\vee h(Y(t)))^{1-\alpha}}\mathrm{d}t \le \frac{|x-y|^{2\varepsilon}}{\varepsilon}. \tag{2.22}$$

Let

$$M(t) = -\int_0^t\frac{\xi(s)}{|X(s)-Y(s)|^{\varepsilon}}\langle\sigma(s)^{-1}(X(s)-Y(s)),\mathrm{d}W(s)\rangle.$$

By (2.22), Hölder's inequality, and $\tau_n\le\tau\le T$, we see that

$$\begin{aligned}
\langle M\rangle(\tau_n) &= \int_0^{\tau_n}\xi(t)^2\frac{\|X(t)-Y(t)\|_{\sigma(t)}^2}{|X(t)-Y(t)|^{2\varepsilon}}\mathrm{d}t \\
&\le \left(\int_0^{\tau_n}\frac{\eta(t)\|X(t)-Y(t)\|_{\sigma(t)}^{\theta}\mathrm{e}^{-\varepsilon\int_0^t\gamma(s)\mathrm{d}s}}{|X(t)-Y(t)|^{\varepsilon\theta}(h(X(t))\vee h(Y(t)))^{1-\alpha}}\mathrm{d}t\right)^{\frac{2}{\theta}} \\
&\quad\times\left(\int_0^{\tau_n}\frac{\xi(t)^{\frac{2\theta}{\theta-2}}(h(X(t))\vee h(Y(t)))^{\frac{2(1-\alpha)}{\theta-2}}\mathrm{e}^{\frac{2\varepsilon}{\theta-2}\int_0^t\gamma(s)\mathrm{d}s}}{\eta(t)^{\frac{2}{\theta-2}}}\mathrm{d}t\right)^{\frac{\theta-2}{\theta}} \\
&\le \frac{|x-y|^{\frac{4\varepsilon}{\theta}}}{\varepsilon^{\frac{2}{\theta}}}\left(\int_0^{\tau_n}\frac{\xi(t)^{\frac{2\theta}{\theta-2}}(h(X(t))\vee h(Y(t)))^{\frac{2(1-\alpha)}{\theta-2}}\mathrm{e}^{\frac{2\varepsilon}{\theta-2}\int_0^t\gamma(s)\mathrm{d}s}}{\eta(t)^{\frac{2}{\theta-2}}}\mathrm{d}t\right)^{\frac{\theta-2}{\theta}}
\end{aligned} \tag{2.23}$$

is bounded. Then, by Girsanov's theorem, $\mathrm{d}\mathbb{Q}_n := R_n\mathrm{d}\mathbb{P}$ is a probability measure under which

$$\tilde{W}_n(t) := W(t) + \int_0^{t\wedge\tau_n}\frac{\xi(s)}{|X(s)-Y(s)|^{\varepsilon}}\sigma(s)^{-1}(X(s)-Y(s))\mathrm{d}s$$

is a cylindrical Brownian motion.

By $\varepsilon=\frac{\theta}{\theta+2}$ and $\frac{2(1-\alpha)}{\theta-2}\le\alpha+1$, (2.23) and Hölder's inequality yield

$$\langle M\rangle(\tau_n) \le \left(\frac{\theta+2}{\theta}\right)^{\frac{2}{\theta}} |x-y|^{\frac{4}{\theta+2}} \tilde{C}(T) \left(\int_0^{\tau_n} \big(h(X(t)) \vee h(Y(t))\big)^{\alpha+1} \mathrm{d}t\right)^{\frac{2(1-\alpha)}{\theta(\alpha+1)}},$$

where due to the definition of ξ and (2.21),

$$\begin{aligned}
\tilde{C}(T) &:= \left(\int_0^T \frac{\xi(t)^{\frac{2\theta(\alpha+1)}{\theta(\alpha+1)-4}} \mathrm{e}^{\frac{2\varepsilon(\alpha+1)}{\theta(\alpha+1)-4}\int_0^t \gamma(s)\mathrm{d}s}}{\eta(t)^{\frac{2(\alpha+1)}{\theta(\alpha+1)-4}}} \mathrm{d}t\right)^{\frac{\theta(\alpha+1)-4}{\theta(\alpha+1)}} \\
&= \left(\frac{\theta+2}{\theta}\right)^2 c_1(\theta,T)^{-\frac{\theta(\alpha+1)+4}{\theta(\alpha+1)}} |x-y|^{\frac{2\theta}{\theta+2}}.
\end{aligned}$$

Therefore,

$$\begin{aligned}
\langle M\rangle(\tau_n) \le\ & \left(\frac{\theta+2}{\theta}\right)^{\frac{2(\theta+1)}{\theta}} c_1(\theta,T)^{-\frac{\theta(\alpha+1)+4}{\theta(\alpha+1)}} |x-y|^2 \qquad (2.24)\\
& \times \left(\int_0^{\tau_n} \big(h(X(t)) \vee h(Y(t))\big)^{\alpha+1} \mathrm{d}t\right)^{\frac{2(1-\alpha)}{\theta(\alpha+1)}}.
\end{aligned}$$

Finally, we have

$$\mathbb{E}\{R_n \log R_n\} = \mathbb{E}_{\mathbb{Q}_n} \log R_n = \frac{1}{2}\mathbb{E}_{\mathbb{Q}}\langle M\rangle(\tau_n),$$

and for $p>1$ and $\tilde{M}_n$ defined as M using the $\mathbb{Q}_n$-cylindrical Brownian motion $\tilde{W}_n$ to replace $W(t)$ (note that $\langle M\rangle = \langle \tilde{M}_n\rangle$), we have

$$\begin{aligned}
\mathbb{E}R_n^{\frac{p}{p-1}} &= \mathbb{E}_{\mathbb{Q}_n} R_n^{\frac{1}{p-1}} = \mathbb{E}_{\mathbb{Q}} \mathrm{e}^{\frac{1}{p-1}\tilde{M}(\tau_n) + \frac{1}{2(p-1)}\langle M\rangle(\tau_n)} \\
&\le \left(\mathbb{E}_{\mathbb{Q}_n} \mathrm{e}^{\frac{2}{p-1}\tilde{M}(\tau_n) - \frac{2}{(p-1)^2}\langle M\rangle(\tau_n)}\right)^{\frac{1}{2}} \left(\mathbb{E}_{\mathbb{Q}_n} \mathrm{e}^{\frac{p+1}{(p-1)^2}\langle M\rangle(\tau_n)}\right)^{\frac{1}{2}} \\
&= \left(\mathbb{E}_{\mathbb{Q}_n} \mathrm{e}^{\frac{p+1}{(p-1)^2}\langle M\rangle(\tau_n)}\right)^{\frac{1}{2}}.
\end{aligned}$$

Combining these with (2.24) and using Jensen's inequality, we finish the proof. □

Lemma 2.3.3 *Assume* (2.18) *and* (2.19). *Let*

$$\lambda(T) = \frac{1}{8\max_{t\in[0,T]} \|\sigma(t)\|_{HS}^2 \mathrm{e}^{-\int_0^t (\gamma(s)+1)\mathrm{d}s}}, \quad \tilde{\lambda}(T) = \mathrm{e}^{-T}\lambda(T) \min_{t\in[0,T]} \tilde{\delta}(t).$$

Then

$$\mathbb{E}\exp\left[\tilde{\lambda}(T)\int_0^T h(X(t))^{\alpha+1}\mathrm{d}t\right] \le \exp\left[2\lambda(T)|x|^2 + 2\lambda(T)\int_0^T \tilde{q}(t)\mathrm{d}t\right].$$

Proof. By (2.18) and Itô's formula,

$$\begin{aligned}
\mathrm{d}|X(t)|^2 \le\ & \{q(t) - \delta(t)h(X(t))^{\alpha+1} + \gamma(t)|X(t)|^2\}\mathrm{d}t \\
& + 2\langle X(t), \sigma(t)\mathrm{d}W(t)\rangle. \qquad (2.25)
\end{aligned}$$

So

$$\mathrm{d}\{|X(t)|^2\mathrm{e}^{-\int_0^t(\gamma(s)+1)\mathrm{d}s}\} \le \mathrm{e}^{-\int_0^t(\gamma(s)+1)\mathrm{d}s}\{q(t)-|X(t)|^2\}\mathrm{d}t$$
$$+2\mathrm{e}^{-\int_0^t(\gamma(s)+1)\mathrm{d}s}\langle X(t),\sigma(t)\mathrm{d}W(t)\rangle.$$

This and the fact that $\mathbb{E}\mathrm{e}^{M(t)} \le (\mathbb{E}\mathrm{e}^{2\langle M\rangle(t)})^{\frac{1}{2}}$ for a continuous martingale $M(t)$ imply that

$$\mathbb{E}\exp\left[\lambda(T)\int_0^T \mathrm{e}^{-\int_0^t(\gamma(s)+1)\mathrm{d}s}|X(t)|^2\mathrm{d}t-\lambda(T)|x|^2-\lambda(T)\int_0^T\tilde{q}(t)\mathrm{e}^{-t}\mathrm{d}t\right]$$
$$\le\left(\mathbb{E}\exp\left[8\lambda(T)^2\int_0^T\mathrm{e}^{-2\int_0^t(\gamma(s)+1)\mathrm{d}s}\|\sigma(t)\|_{HS}^2|X(t)|^2\mathrm{d}t\right]\right)^{\frac{1}{2}}$$
$$\le\left(\mathbb{E}\exp\left[\lambda(T)\int_0^T\mathrm{e}^{-\int_0^t(\gamma(s)+1)\mathrm{d}s}|X(t)|^2\mathrm{d}t\right]\right)^{\frac{1}{2}}.$$

By an approximation argument, we may assume that the expectations in this display are finite, so that

$$\mathbb{E}\exp\left[\lambda(T)\int_0^T\mathrm{e}^{-\int_0^t(\gamma(s)+1)\mathrm{d}s}|X(t)|^2\mathrm{d}t\right]$$
$$\le\exp\left[2\lambda(T)|x|^2+2\lambda(T)\int_0^T\tilde{q}(t)\mathrm{e}^{-t}\mathrm{d}t\right]. \tag{2.26}$$

Combining (2.26) with (2.25), we obtain

$$\mathbb{E}\exp\left[\tilde{\lambda}(T)\int_0^T h(X(t))^{\alpha+1}\mathrm{d}t\right]\le\mathbb{E}\exp\left[\lambda(T)\mathrm{e}^{-T}\int_0^T\tilde{\delta}(t)h(X(t))^{\alpha+1}\mathrm{d}t\right]$$
$$\le\exp\left[\lambda(T)\mathrm{e}^{-T}|x|^2+\lambda(T)\mathrm{e}^{-T}\int_0^T\tilde{q}(t)\mathrm{d}t\right]$$
$$\times\mathbb{E}\exp\left[2\lambda(T)\mathrm{e}^{-T}\int_0^T\mathrm{e}^{-\int_0^t\gamma(s)\mathrm{d}s}\langle X(t),\sigma(t)\mathrm{d}W(t)\rangle\right]$$
$$\le\exp\left[\lambda(T)|x|^2+\lambda(T)\int_0^T\tilde{q}(t)\mathrm{e}^{-t}\mathrm{d}t\right]$$
$$\times\left(\mathbb{E}\exp\left[\lambda(T)\int_0^T\mathrm{e}^{-\int_0^t(\gamma(s)+1)\mathrm{d}s}|X(t)|^2\mathrm{d}t\right]\right)^{\frac{1}{2}}$$
$$\le\exp\left[2\lambda(T)|x|^2+2\lambda(T)\int_0^T\tilde{q}(t)\mathrm{e}^{-t}\mathrm{d}t\right].$$

□

Proof of Theorem 2.3.1. According to Theorem 1.1.1 and Lemma 2.3.2, we have only to estimate the expectation and the exponential moment of $\int_0^T(h(X(t))\vee h(Y(t)))^{\alpha+1}\mathrm{d}t$ with respect to $\mathrm{d}\mathbb{Q}:=R\mathrm{d}\mathbb{P}$.

(1) By (2.18) and Itô's formula, we have

$$d|Y(t)|^2 \le \{q(t) - \delta(t)h(Y(t))^{\alpha+1} + \gamma(t)|Y(t)|^2\}dt + 2\langle Y(t), \sigma(t)d\tilde{W}_n(t)\rangle, \ t \le \tau_n.$$

Then

$$\mathbb{E}_{\mathbb{Q}_n} \int_0^{\tau_n} h(Y(t))^{\alpha+1}dt \le \frac{|y|^2 + \int_0^T \tilde{q}(t)dt}{\min_{t\in[0,T]} \tilde{\delta}(t)}, \tag{2.27}$$

$$\mathbb{E}_{\mathbb{Q}_n}|Y(t\wedge\tau_n)|^2 \le \left(|y|^2 + \int_0^t \tilde{q}(s)ds\right)e^{\int_0^t \gamma(s)ds}, \ t \in [0,T]. \tag{2.28}$$

Next, by (2.18) and Itô's formula, we have

$$\begin{aligned} d|X(t)|^2 &\le \{q(t) - \delta(t)h(X(t))^{\alpha+1} + \gamma(t)|X(t)|^2\}dt + 2\langle X(t), \sigma(t)dW(t)\rangle \\ &\le \Big\{q(t) + \gamma(t)|X(t)|^2 - \delta(t)h(X(t))^{\alpha+1} - \frac{2\xi(t)\langle X(t), X(t)-Y(t)\rangle}{|X(t)-Y(t)|^\varepsilon}\Big\}dt \\ &\quad + 2\langle X(t), \sigma(t)d\tilde{W}_n(t)\rangle, \quad t \le \tau_n. \end{aligned} \tag{2.29}$$

Noting that

$$-\frac{\langle u, u-v\rangle}{|u-v|^\varepsilon} \le |v|\cdot|u-v|^{1-\varepsilon} - |u-v|^{2-\varepsilon} \le \frac{|v|^{2-\varepsilon}}{1-\varepsilon}\left(\frac{1-\varepsilon}{2-\varepsilon}\right)^{2-\varepsilon},$$

we obtain

$$\begin{aligned} d\{|X(t)|^2 e^{-\int_0^t \gamma(s)ds}\} &\le 2e^{-\int_0^t \gamma(s)ds}\langle X(t), \sigma(t)d\tilde{W}_n(t)\rangle \\ &+ e^{-\int_0^t \gamma(s)ds}\Big\{q(t) - \delta(t)h(X(t))^{\alpha+1} + \frac{2\xi(t)|Y(t)|^{2-\varepsilon}}{1-\varepsilon}\left(\frac{1-\varepsilon}{2-\varepsilon}\right)^{2-\varepsilon}\Big\}dt, \ t \le \tau_n. \end{aligned}$$

Combining this with (2.28) and (2.20), we arrive at

$$\begin{aligned} &\mathbb{E}_{\mathbb{Q}_n} \int_0^{\tau_n} h(X(t))^{\alpha+1}dt \\ &\le \frac{|x|^2 + \int_0^T e^{-\int_0^t \gamma(s)ds}\{q(t) + \frac{2\xi(t)}{1-\varepsilon}(\frac{1-\varepsilon}{2-\varepsilon})^{2-\varepsilon}(\mathbb{E}_{\mathbb{Q}_n}|Y(t\wedge\tau_n)|^2)^{\frac{2-\varepsilon}{2}}\}dt}{\min_{t\in[0,T]}\delta(t)e^{-\int_0^t \gamma(s)ds}} \\ &\le \frac{|x|^2 + \int_0^T \tilde{q}(t)dt + \frac{2|x-y|^\varepsilon}{\varepsilon(1-\varepsilon)}(\frac{1-\varepsilon}{2-\varepsilon})^{2-\varepsilon}(|y|^2 + \int_0^T \tilde{q}(t)dt)^{\frac{2-\varepsilon}{2}}}{\min_{[0,T]}\tilde{\delta}}. \end{aligned}$$

Combining this with (2.27), we conclude that

$$\begin{aligned} &\mathbb{E}_{\mathbb{Q}_n} \int_0^{\tau_n} \left(h(X(t)) \vee h(Y(t))\right)^{\alpha+1}dt \\ &\le \frac{|x|^2 + |y|^2 + 2\int_0^T \tilde{q}(t)dt + \frac{2|x-y|^\varepsilon}{\varepsilon(1-\varepsilon)}(\frac{1-\varepsilon}{2-\varepsilon})^{2-\varepsilon}(|y|^2 + \int_0^T \tilde{q}(t)dt)^{\frac{2-\varepsilon}{2}}}{\min_{t\in[0,T]}\tilde{\delta}(t)} \\ &= c_2(\theta, T, x, y). \end{aligned}$$

Thus, by Lemma 2.3.2, we obtain

$$\mathbb{E}\{R_n \log R_n\} \leq \frac{|x-y|^2}{2}\Big(\frac{\theta+2}{\theta}\Big)^{\frac{2(\theta+1)}{\theta}} \frac{c_2(\theta,T,x,y)^{\frac{2(1-\alpha)}{\theta(\alpha+1)}}}{c_1(\theta,T)^{\frac{\theta(\alpha+1)+4}{\theta(\alpha+1)}}}.$$

Letting $n \uparrow \infty$, we conclude that R given in (2.10) is a probability density with

$$\mathbb{E}\{R \log R\} \leq \frac{|x-y|^2}{2}\Big(\frac{\theta+2}{\theta}\Big)^{\frac{2(\theta+1)}{\theta}} \frac{c_2(\theta,T,x,y)^{\frac{2(1-\alpha)}{\theta(\alpha+1)}}}{c_1(\theta,T)^{\frac{\theta(\alpha+1)+4}{\theta(\alpha+1)}}}.$$

Let $\mathrm{d}\mathbb{Q} = R\mathrm{d}\mathbb{P}$. Then the first log-Harnack inequality in (1) follows from Theorem 1.1.1. As for the second inequality, by the Markov property and Jensen's inequality, we need consider only $T \in (0,1]$. In this case, the second inequality follows from the first, since for some constants $C_1, C_2, C_3 > 0$, we have $c_1(\theta,T) \geq C_1 T$ and

$$c_2(\theta,T,x,y) \leq C_2\Big\{|x|^2+|y|^2+T+|x-y|^{\frac{\theta}{\theta+2}}\left(|y|^2+T\right)^{\frac{\theta+4}{2(\theta+2)}}\Big\} \leq C_3\Big\{|x|^2+|y|^2+T\Big\}.$$

(2) Again we need to prove the result only for $T \in (0,1]$. According to Theorem 1.1.1, it suffices to find a constant $C > 0$ such that for every $T \in (0,1]$, $p > 1$, and $x,y \in \mathbb{H}$,

$$\begin{aligned}&(\mathbb{E}R^{\frac{p}{p-1}})^{p-1}\\ &\leq \exp\left[C\left(\frac{p|x-y|^2(1+|x|^2+|y|^2)}{(p-1)T^{\frac{\theta(\alpha+1)+4}{\theta(\alpha+1)}}} + \frac{(\frac{p}{p-1})^{\frac{\alpha\theta+\theta+2-2\alpha}{\alpha(\theta+2)+\theta-2}}|x-y|^{\frac{2\theta(\alpha+1)}{\alpha(\theta+2)+\theta-2}}}{T^{\frac{\theta(\alpha+1)+4}{\alpha(\theta+2)+\theta-2}}}\right)\right].\end{aligned} \tag{2.30}$$

By Lemma 2.3.3 with $n \uparrow \infty$ and noting that the distribution of $Y(t)$ under $\mathbb{Q}$ coincides with that of $X^y(t)$ under $\mathbb{P}$, there exists a constant $C_1 > 0$ such that for every $r > 0$ and $s,T \in (0,1]$,

$$\begin{aligned}&\mathbb{E}_{\mathbb{Q}}\exp\left[r\left(\int_0^T h(Y(t))^{\alpha+1}\mathrm{d}t\right)^{\frac{2(1-\alpha)}{\theta(\alpha+1)}}\right]\\ &\leq \mathbb{E}_{\mathbb{Q}}\exp\left[s\tilde{\lambda}(T)\int_0^T h(Y(t))^{\alpha+1}\mathrm{d}t + C_1 r^{\frac{\theta(1+\alpha)}{\alpha(\theta+2)+\theta-2}}(\tilde{\lambda}(T)s)^{-\frac{2(1-\alpha)}{\alpha(\theta+2)+\theta-2}}\right]\\ &\leq \exp\left[C_1 s(|y|^2+1) + C_1 r^{\frac{\theta(1+\alpha)}{\alpha(\theta+2)+\theta-2}}(\tilde{\lambda}(T)s)^{-\frac{2(1-\alpha)}{\alpha(\theta+2)+\theta-2}}\right].\end{aligned}$$

Taking $s = 1 \wedge r$, we see that $\inf_{T\in(0,1]}\tilde{\lambda}(T) > 0$ implies

$$r^{\frac{\theta(1+\alpha)}{\alpha(\theta+2)+\theta-2}}(\tilde{\lambda}(T)s)^{-\frac{2(1-\alpha)}{\alpha(\theta+2)+\theta-2}} \leq C'\left(r + r^{\frac{\theta(1+\alpha)}{\alpha(\theta+2)+\theta-2}}\right)$$

for some constant $C' > 0$ and all $r > 0$, $T \in (0,1]$. Hence,

$$\mathbb{E}_{\mathbb{Q}} \exp\left[r\left(\int_0^T h(Y(t))^{\alpha+1} dt\right)^{\frac{2(1-\alpha)}{\theta(\alpha+1)}}\right] \le \exp\left[C_2\left(r + r|y|^2 + r^{\frac{\theta(\alpha+1)}{\alpha(\theta+2)+\theta-2}}\right)\right]$$

holds for some constant $C_2 > 0$ and all $r > 0$, $T \in (0,1]$. Consequently, for every constant $r > 0$ (recall that the distribution of Y under $\mathbb{Q}$ coincides with that of X^y under $\mathbb{P}$),

$$\begin{aligned} &\mathbb{E}_{\mathbb{Q}} \exp\left[\frac{r|x-y|^2}{T^{\frac{\theta(\alpha+1)+4}{\theta(\alpha+1)}}}\left(\int_0^T h(Y(t))^{\alpha+1} dt\right)^{\frac{2(1-\alpha)}{\theta(\alpha+1)}}\right] \\ &\le \exp\left[\frac{C_2 r|x-y|^2(|y|^2+1)}{T^{\frac{\theta(\alpha+1)+4}{\theta(\alpha+1)}}} + C_2\left(\frac{r|x-y|^2}{T^{\frac{\theta(\alpha+1)+4}{\theta(\alpha+1)}}}\right)^{\frac{\theta(\alpha+1)}{\alpha(\theta+2)+\theta-2}}\right], \end{aligned} \tag{2.31}$$

$$\begin{aligned} &\mathbb{E} \exp\left[\frac{r|x-y|^2}{T^{\frac{\theta(\alpha+1)+4}{\theta(\alpha+1)}}}\left(\int_0^T h(X(t))^{\alpha+1} dt\right)^{\frac{2(1-\alpha)}{\theta(\alpha+1)}}\right] \\ &\le \exp\left[\frac{C_2 r|x-y|^2(|x|^2+1)}{T^{\frac{\theta(\alpha+1)+4}{\theta(\alpha+1)}}} + C_2\left(\frac{r|x-y|^2}{T^{\frac{\theta(\alpha+1)+4}{\theta(\alpha+1)}}}\right)^{\frac{\theta(\alpha+1)}{\alpha(\theta+2)+\theta-2}}\right]. \end{aligned} \tag{2.32}$$

Combining these with Lemma 2.3.2 and noting that $\frac{p+1}{p} \le 2$, we obtain for some constants $c, C_3 > 0$ that

$$\begin{aligned} (\mathbb{E}R^{\frac{p}{p-1}})^4 &\le \left(\mathbb{E}_{\mathbb{Q}} \exp\left[\frac{cp|x-y|^2}{(p-1)^2 T^{\frac{\theta(\alpha+1)+4}{\theta(\alpha+1)}}}\left(\int_0^T \{h(X(t)) \vee h(Y(t))\}^{\alpha+1} dt\right)^{\frac{2(1-\alpha)}{\theta(\alpha+1)}}\right]\right)^2 \\ &\le \left(\mathbb{E}_{\mathbb{Q}} \exp\left[\frac{2cp|x-y|^2}{(p-1)^2 T^{\frac{\theta(\alpha+1)+4}{\theta(\alpha+1)}}}\left(\int_0^T h(X(t))^{\alpha+1} dt\right)^{\frac{2(1-\alpha)}{\theta(\alpha+1)}}\right]\right) \\ &\quad \times \left(\mathbb{E}_{\mathbb{Q}} \exp\left[\frac{2cp|x-y|^2}{(p-1)^2 T^{\frac{\theta(\alpha+1)+4}{\theta(\alpha+1)}}}\left(\int_0^T h(Y(t))^{\alpha+1} dt\right)^{\frac{2(1-\alpha)}{\theta(\alpha+1)}}\right]\right) \\ &\le (\mathbb{E}R^{\frac{p}{p-1}})^{\frac{p-1}{p}} \left(\mathbb{E} \exp\left[\frac{2cp^2|x-y|^2}{(p-1)^2 T^{\frac{\theta(\alpha+1)+4}{\theta(\alpha+1)}}}\left(\int_0^T h(X(t))^{\alpha+1} dt\right)^{\frac{2(1-\alpha)}{\theta(\alpha+1)}}\right]\right)^{\frac{1}{p}} \\ &\quad \times \exp\left[\frac{2C_2 cp|x-y|^2(|y|^2+1)}{(p-1)^2 T^{\frac{\theta(\alpha+1)+4}{\theta(\alpha+1)}}} + C_2\left(\frac{2cp|x-y|^2}{(p-1)^2 T^{\frac{\theta(\alpha+1)+4}{\theta(\alpha+1)}}}\right)^{\frac{\theta(\alpha+1)}{\alpha(\theta+2)+\theta-2}}\right] \\ &\le (\mathbb{E}R^{\frac{p}{p-1}})^{\frac{p-1}{p}} \\ &\quad \times \exp\left[C_3\left(\frac{p|x-y|^2(1+|x|^2+|y|^2)}{(p-1)^2 T^{\frac{\theta(\alpha+1)+4}{\theta(\alpha+1)}}} + \frac{(\frac{p}{p-1})^{\frac{\alpha\theta+\theta+2-2\alpha}{\alpha(\theta+2)+\theta-2}} |x-y|^{\frac{2\theta(\alpha+1)}{\alpha(\theta+2)+\theta-2}}}{(p-1) T^{\frac{\theta(\alpha+1)+4}{\alpha(\theta+2)+\theta-2}}}\right)\right]. \end{aligned}$$

This implies (2.30) for some constant $C > 0$. $\square$

2.4 Applications to Specific Models

In this section we apply Theorems 2.2.1, 2.3.1 and Corollary 2.2.4 to specific models presented in Sect. 2.1.

2.4.1 Stochastic Generalized Porous Media Equations

Let $(E,\mathscr{B},\mathbf{m})$ be a separable probability space and $(L,\mathscr{D}(L))$ a negative definite self-adjoint linear operator on $L^2(\mathbf{m})$ having discrete spectrum. Let

$$(0<)\lambda_1 \le \lambda_2 \le \cdots$$

be all the eigenvalues of $-L$ including multiplicities with unit eigenfunctions $\{e_i\}_{i\ge 1}$.

Let $\mathbb{H}$ be the dual space of $\mathscr{D}((-L)^{\frac{1}{2}})$ with respect to $L^2(\mathbf{m})$; i.e., $\mathbb{H}$ is the completion of $L^2(\mathbf{m})$ under the inner product

$$\langle x,y\rangle := \sum_{i=1}^{\infty} \frac{1}{\lambda_i}\langle x,e_i\rangle_2 \langle y,e_i\rangle_2.$$

Recall that $\langle\ ,\ \rangle_2$ is the inner product in $L^2(\mathbf{m})$. Let $W(t)$ be the cylindrical Brownian motion on $\mathbb{H}$ with respect to a complete filtered probability space $(\Omega,\{\mathscr{F}_t\}_{t\ge 0},\mathbb{P})$. Let

$$\Psi,\Phi : [0,\infty)\times\mathbb{R}\to\mathbb{R}$$

be measurable and continuous in the second variable, and let

$$\sigma : [0,\infty)\to \mathscr{L}_{HS}(\mathbb{H})$$

be measurable such that $\|\sigma\|_{HS}\in L^2_{loc}([0,\infty);\mathrm{d}t)$. We consider the equation

$$\mathrm{d}X(t) = \{L\Psi(t,X(t)) + \Phi(t,X(t))\}\mathrm{d}t + \sigma(t)\mathrm{d}W(t). \tag{2.33}$$

Let $\alpha\ge 1$ be a fixed number and assume that L^{-1} is bounded in $L^{\alpha+1}(\mathbf{m})$ if $\Phi\ne 0$. It is easy to see that this assumption holds if L is a Dirichlet operator. To ensure the existence and uniqueness of the solution to (2.33), we assume (2.2) and (2.3) as in Example 2.1.1.

Theorem 2.4.1 *Assume* $\|\sigma(\cdot)\|_{HS}\in L^2_{loc}([0,\infty);\mathrm{d}t)$, (2.2), *and* (2.3). *If there exists a nonnegative constant* $\theta\in[2,\infty)\cap(\alpha-1,\infty)$ *such that*

$$\|x\|^{\alpha+1}_{L^{\alpha+1}(\mathbf{m})} \ge \xi(t)\|x\|^{\theta}_{\sigma(t)}|x|^{\alpha+1-\theta},\quad x\in L^{\alpha+1}(\mathbf{m}),\ t\ge 0, \tag{2.34}$$

for some strictly positive function $\xi\in C([0,\infty))$, *then:*

(1) *The Harnack inequalities in Theorem 2.2.1 hold for* $\eta(t) := \delta(t)\xi(t)$.
(2) *When the equation is time-homogeneous, i.e.,* $\sigma(t)$, $\Psi(t,\cdot)$, *and* $\Phi(t,\cdot)$ *are independent of* t *such that* δ, γ, *and* ξ *are constants, and either* $\alpha > 1$ *or* $\alpha = 1$ *but* $\delta\lambda_1 > \gamma$, *then* P_t *has a unique invariant probability measure* μ *that has full support such that (2.13) and the assertions in Corollary 2.2.4 hold.*

Proof. Simply note that (2.3) and (2.34) imply (2.5) for $\eta(t) = \delta(t)\xi(t)$, and since $\|\cdot\|^2_{L^2(\mathbf{m})} \geq \lambda_1|\cdot|^2$, $\alpha > 1$ or $\alpha = 1$ but $\delta\lambda_1 > \gamma$ implies (2.12) for some $C \geq 0$ and $\delta > 0$. □

It is easy to construct examples of Φ and Ψ such that condition (2.3) holds. Below we present a simple example to illustrate condition (2.34) and that $\sigma \in \mathscr{L}_{HS}(\mathbb{H})$, so that Theorem 2.4.1 applies.

Example 2.4.1 Let $\sigma e_i = \sigma_i e_i$, $i \geq 1$, for a sequence $\{\sigma_i\}_{i\geq 1} \subset \mathbb{R}$. If

$$\sum_{i=1}^{\infty} \sigma_i^2 < \infty, \quad \inf_{i\geq 1}\{\lambda_i\sigma_i^2\} > 0, \tag{2.35}$$

then $\sigma \in \mathscr{L}_{HS}(\mathbb{H})$ and (2.34) holds for some constant $\xi > 0$. Indeed, the first inequality implies that $\sigma \in \mathscr{L}_{HS}(\mathbb{H})$, while the second inequality implies (2.34) for some $\xi > 0$, since

$$\|\cdot\|^2_{L^{\alpha+1}(\mathbf{m})} \geq \|\cdot\|^2_{L^2(\mathbf{m})} \geq \inf_{i\geq 1}(\sigma_i^2\lambda_i)\|\cdot\|^2_{\sigma}.$$

In particular, (2.35) holds if $\sum_{i=1}^{\infty}\frac{1}{\lambda_i} < \infty$ and

$$\frac{1}{C\lambda_i} \leq \sigma_i^2 \leq \frac{C}{\lambda_i}, \quad i \geq 1,$$

holds for some constant $C \geq 1$. When $\sum_{i\geq 1}\frac{1}{\lambda_i} < \infty$, one may easily choose $(\sigma_i)_{i\geq 1}$ satisfying (2.35). That is the case for $L = \Delta$ the Dirichlet Laplacian on a bounded interval.

2.4.2 Stochastic p-Laplacian Equations

We simply consider the equation in Example 2.1.2 for $p \geq 2$, $d = 1$, and $\mathbf{m}(\mathrm{d}x) = \mathrm{d}x$ on $(0,1)$. In this case, we have $\mathbb{H} = L^2(\mathbf{m})$ and $\mathbb{V} = \mathbb{H}_0^{1,p}$. Let Δ be the Dirichlet Laplacian on $(0,1)$. Then $\{(\pi i)^2\}_{i\geq 1}$ are all eigenvalues of $-\Delta$ with unit eigenfunctions $e_i(x) := \sqrt{2}\sin(i\pi x)$. Assume that there exist a sequence of constants $(\sigma_i)_{i\geq 1}$ and a constant $c > 0$ such that

$$\sigma e_i = \sigma_i e_i, \quad \sigma_i^2 \geq \frac{c}{i^4}, \quad \sum_{i\geq 1}\sigma_i^2 < \infty. \tag{2.36}$$

Then $b(t,u) = \mathrm{div}(|\nabla u|^{p-2}\nabla u)$. According to Example 3.3 in [27], we have

$$2_{\mathbb{V}^*}\langle b(t,u) - b(t,v), u-v\rangle_{\mathbb{V}} \le -2^{p-1}\mathbf{m}(|\nabla(u-v)|^2)^{\frac{p}{2}}.$$

Since by (2.36),

$$\mathbf{m}(|\nabla(u-v)|^2) = \sum_{i\ge 1}\pi^{2i}\mathbf{m}((u-v)e_i)^2 \ge c\|u-v\|_\sigma^2,$$

we obtain

$$2_{\mathbb{V}^*}\langle b(t,u) - b(t,v), u-v\rangle_{\mathbb{V}} \le -2^{p-1}c^{\frac{p}{2}}\|u-v\|_\sigma^p.$$

So the assertion in Theorem 2.2.1 holds for $\theta = p$, $\alpha = p-1$, $\eta(t) = 2^{p-1}c^{\frac{p}{2}}$, and $\gamma(t) = 0$. Therefore, there exists a constant $C > 0$ such that for every positive $f \in \mathscr{B}_b(\mathbb{H})$, $x, y \in \mathbb{H}$, and $T > 0$, we have

$$P_T \log f(y) \le \log P_T f(x) + \frac{C|x-y|^{\frac{4}{p}}}{T^{\frac{p+2}{p}}},$$

and for every $p' > 1$,

$$(P_T f(y))^{p'} \le (P_T f^{p'}(x)) \exp\left[\frac{p'C|x-y|^{\frac{4}{p}}}{(p'-1)T^{\frac{p+2}{p}}}\right].$$

2.4.3 Stochastic Generalized Fast-Diffusion Equations

Let $(E,\mathscr{B},\mathbf{m}), (L,\mathscr{D}(L))$, $\mathbb{H}$, σ, and $W(t)$ be as in Sect. 2.4.1. We consider the equation

$$\mathrm{d}X(t) = \Big\{L\Psi(t,X(t)) + \beta(t)X(t)\Big\}\mathrm{d}t + \sigma(t)\mathrm{d}W(t), \tag{2.37}$$

where $\beta \in C([0,\infty))$, $\Psi : [0,\infty)\times\mathbb{R} \to \mathbb{R}$ is measurable, continuous in the second variable, such that for some constant $\alpha \in (0,1)$ and strictly positive $\zeta, \delta \in C([0,\infty))$,

$$2\big(\Psi(t,s_1) - \Psi(t,s_2)\big)(s_1 - s_2) \ge \zeta(t)|s_1 - s_2|^2(|s_1| \vee |s_2|)^{\alpha-1}, \quad s_1, s_2 \in \mathbb{R}, t \ge 0, \tag{2.38}$$

$$s\Psi(t,s) \ge \delta(t)|s|^{\alpha+1}, \quad \sup_{t\in[0,T], s\ge 0}\frac{|\Psi(t,s)|}{1+|s|^\alpha} < \infty, \quad s \in \mathbb{R}, t \ge 0. \tag{2.39}$$

By the mean value theorem and $\alpha < 1$, one has $(s_1 - s_2)(s_1^\alpha - s_2^\alpha) \ge \alpha|s_1 - s_2|^2(|s_1| \vee |s_2|)^{\alpha-1}$. So a simple example of Ψ for these conditions to hold is $\Psi(t,s) = s^\alpha := |s|^\alpha \mathrm{sgn}(s)$, for which (2.37) is known as the stochastic fast-diffusion equation with linear drift.

Let $\mathbb{V} = L^{\alpha+1}(\mathbf{m}) \cap \mathbb{H}$. It is easy to see that **(A2.1)–(A2.4)** hold (see [40, Theorem 3.9] for a more general result). Let $X^x(t)$ be the unique solution for $X(0) = x \in \mathbb{H}$, and let P_t be the associated Markov operator.

Theorem 2.4.2 *Assume that* (2.38) *and* (2.39) *hold for some strictly positive functions* $\zeta, \delta \in C([0,\infty))$. *If there exist a constant* $\theta \geq \frac{4}{\alpha+1}$ *and a strictly positive function* $\xi \in C([0,\infty))$ *such that*

$$\|u\|_{L^{\alpha+1}(\mathbf{m})}^2 \cdot |u|^{\theta-2} \geq \xi(t)\|u\|_{\sigma(t)}^{\theta}, \quad u \in L^{\alpha+1}(\mathbf{m}),\ t \geq 0, \tag{2.40}$$

then the assertions in Theorem 2.3.1 hold for

$$\gamma(t) := 2\beta(t), \quad q(t) := \|\sigma(t)\|_{HS}^2, \quad \eta(t) := 2^{\frac{\alpha-1}{1+\alpha}}\zeta(t)\xi(t).$$

Proof. Let $h(u) := \|u\|_{L^{\alpha+1}(\mathbf{m})}$. Obviously, (2.37) and the first inequality in (2.39) imply (2.18) for $q(t) = \|\sigma(t)\|_{HS}^2$ and $\gamma(t) = 2\beta(t)$. It remains to verify (2.19) for the above η and h. By Hölder's inequality, we have

$$\begin{aligned}
\|u-v\|_{L^{\alpha+1}(\mathbf{m})}^{\alpha+1} &= \mathbf{m}(|u-v|^{\alpha+1}) \leq \mathbf{m}\big(|u-v|^2(|u|\vee|v|)^{\alpha-1}\big)^{\frac{\alpha+1}{2}}\mathbf{m}\big((|u|\vee|v|)^{\alpha+1}\big)^{\frac{1-\alpha}{2}}\\
&\leq 2^{\frac{1-\alpha}{2}}\mathbf{m}\big(|u-v|^2(|u|\vee|v|)^{\alpha-1}\big)^{\frac{\alpha+1}{2}}(h(u)\vee h(v))^{\frac{1-\alpha^2}{2}}.
\end{aligned}$$

Combining this with (2.38) and (2.40), we prove (2.19) for $\eta(t) = 2^{\frac{\alpha-1}{1+\alpha}}\zeta(t)\xi(t)$. □

To conclude this subsection, we consider the stochastic fast-diffusion equation for which $\Psi(t,s) = s^\alpha$ for some $\alpha \in (0,1)$.

Corollary 2.4.3 *Consider* (2.37) *for* $\Psi(t,s) = s^\alpha := |s|^\alpha \mathrm{sgn}(s)$. *Let* $(-L, \mathscr{D}(L))$ *be a nonnegative definite self-adjoint operator on* $L^2(\mathbf{m})$ *with discrete spectrum* $(0 <)\lambda_1 \leq \lambda_2 \leq \cdots \leq \lambda_n \uparrow \infty$, *counting multiplicities. Let* $\{e_n\}_{n\geq 1}$ *be the corresponding eigenvectors, which form an orthonormal basis of* $L^2(\mathbf{m})$. *Let* $\sigma(t) = \sigma$ *be such that*

$$\sigma e_i = \sigma_i e_i, \quad i \geq 1,$$

for some constants $\{\sigma_i\}_{i\geq 1}$ *satisfying*

$$\|\sigma\|_{HS}^2 = \sum_{i=1}^{\infty}\sigma_i^2 < \infty. \tag{2.41}$$

If there exist constants $\varepsilon \in (0,1), \theta \geq \frac{4}{\alpha+1}$ *and* $C_1, C_2 > 0$ *such that*

$$|\sigma_i| \geq C_1 \lambda_i^{-\frac{1-\varepsilon}{\theta}}, \quad i \geq 1, \tag{2.42}$$

and the Nash inequality

$$\|f\|_{L^2(\mathbf{m})}^{2+\frac{4}{d}} \leq -C_2\mathbf{m}(fLf), \quad f \in \mathscr{D}(L),\ \mathbf{m}(|f|) = 1, \tag{2.43}$$

holds for some $d \in (0, \frac{2\varepsilon(\alpha+1)}{1-\alpha})$, *then the assertions in Theorem 2.3.1 hold for*

$$\gamma(t) = 2\beta(t), \;\; q(t) = \|\sigma\|_{HS}^2, \;\; \delta(t) = 1, \;\; \text{and } \eta(t) = c,$$

for some constant $c > 0$. *In particular, if* $\beta \leq 0$, *we may take* $\gamma = 0$ *such that the log-Harnack inequality*

$$P_T \log f(y) - \log P_T f(x) \leq \frac{C|x-y|^2(|x|^2+|y|^2+T)^{\frac{2(1-\alpha)}{\theta(1+\alpha)}}}{T^{\frac{\theta(\alpha+1)+4}{\theta(\alpha+1)}}}$$

holds for some constant $C > 0$ *and all* $T > 0$, $x, y \in \mathbb{H}$, *and positive* $f \in \mathscr{B}_b(\mathbb{H})$.

Proof. By Jensen's inequality, it suffices to prove the result for $T \in (0,1]$. Obviously, (2.38) and (2.39) hold for $\zeta(t) = 2\alpha$ and $\delta(t) = 1$. To apply Theorem 2.4.2, it remains to verify (2.40). By (2.42), we have

$$\begin{aligned}\|x\|_\sigma^\theta &= \Big(\sum_{i\geq 1}\frac{\mathbf{m}(xe_i)^2}{\sigma_i^2\lambda_i}\Big)^{\frac{\theta}{2}} \leq \Big(\sum_{i\geq 1}\frac{\mathbf{m}(xe_i)^2}{|\sigma_i|^\theta\lambda_i}\Big)\Big(\sum_{i\geq 1}\frac{\mathbf{m}(xe_i)^2}{\lambda_i}\Big)^{\frac{\theta-2}{2}}\\ &\leq C_1^{-\theta}|x|^{\theta-2}\Big(\sum_{i\geq 1}\frac{\mathbf{m}(xe_i)^2}{\lambda_i^\varepsilon}\Big).\end{aligned}$$

According to the proof of Corollary 3.2 in [31], (2.43) for some $d \in (0, \frac{2\varepsilon(\alpha+1)}{1-\alpha})$ implies that

$$\|x\|_{\alpha+1}^2 \geq c\sum_{i\geq 1}\frac{\mathbf{m}(xe_i)^2}{\lambda_i^\varepsilon}$$

holds for some constant $c > 0$. Therefore, (2.40) holds for $\xi(t) = c'$ for some constant $c' > 0$. Combining these with (2.41) and using Theorem 2.4.2, we may apply Theorem 2.3.1 to

$$\gamma(t) = 2\beta(t), \;\; q(t) = \|\sigma\|_{HS}^2, \;\; \delta(t) = 1, \;\; \text{and } \eta(t) = c,$$

for some constant $c > 0$. Finally, if $\gamma = 0$, then it is easy to see that for some constants $C_1', C_2' > 0$,

$$c_1(\theta,T) \geq C_1'T, \;\; c_2(\theta,T,x,y) \leq C_2'(|x|^2+|y|^2+T), \quad T>0, x,y \in \mathbb{H},$$

which implies the desired log-Harnack inequality according to Theorem 2.3.1(1). □

Example 2.4.2 *Let* $\Psi(t,s) = s^\alpha$ *for some* $\alpha \in (\frac{1}{3}, 1)$, *and let* $L = \Delta$ *be the Dirichlet Laplacian on the open interval* $(0,\pi)$. *Then* $\lambda_i = i^2$ *and the Nash inequality* (2.43) *holds for* $d = 1$. *For every* $\theta \in (\frac{4}{\alpha+1}, \frac{6\alpha+2}{\alpha+1})$ *and* $\varepsilon \in (\frac{1-\alpha}{2(\alpha+1)}, 1-\frac{\theta}{4})$, *we have* $d = 1 \in (0, \frac{2\varepsilon(\alpha+1)}{1-\alpha})$, *as required. So taking* $\sigma_i = i^{-\frac{2(1-\varepsilon)}{\theta}}$, *we see that* (2.41) *and* (2.42) *are satisfied. Therefore, the assertions in Corollary 2.4.3 hold.*

Chapter 3
Semilinear Stochastic Partial Differential Equations

3.1 Mild Solutions and Finite-Dimensional Approximations

Let $(\mathbb{H}, \langle\cdot,\cdot\rangle, |\cdot|)$ be a separable Hilbert space, and let $\tilde{\mathbb{H}}$ be a larger Hilbert space into which $\mathbb{H}$ is densely and continuously embedded. Let $(A, \mathscr{D}(A))$ be a negative definite self-adjoint operator on $\mathbb{H}$ generating a C_0 contraction semigroup with $S(t) = e^{At}$, $t \geq 0$. Let $\mathscr{L}_S(\mathbb{H})$ be the set of all densely defined closed linear operators $(L, \mathscr{D}(L))$ on $\mathbb{H}$ such that for every $s > 0$, $S(s)L$ extends to a unique Hilbert–Schmidt operator on $\mathbb{H}$, which is again denoted by $S(s)L$. We equip $\mathscr{L}_S(\mathbb{H})$ with the σ-field induced by $\{L \mapsto \langle (S(s)L)x, y\rangle : \ s > 0, x, y \in \mathbb{H}\}$. Let $T > 0$ be fixed, and let

$$b : [0,T] \times \mathbb{H} \to \tilde{\mathbb{H}}, \ \ \sigma : [0,T] \times \mathbb{H} \to \mathscr{L}_S(\mathbb{H})$$

be measurable maps. We shall set $|v| = \infty$ for $v \notin \mathbb{H}$. Consider the following SPDE on $\mathbb{H}$:

$$\mathrm{d}X(t) = \big\{AX(t) + b(t, X(t))\big\}\mathrm{d}t + \sigma(t, X(t))\mathrm{d}W(t), \ \ t \in [0,T], \tag{3.1}$$

where $W(t)$ is a cylindrical Brownian motion on $\mathbb{H}$ with respect to a complete filtered probability space $(\Omega, \{\mathscr{F}_t\}_{t\geq 0}, \mathbb{P})$. See [10, 36, 72, 73] for equations with a time-dependent linear operator and a multivalued nonlinear term.

Definition 3.1.1 *An $\mathbb{H}$-valued progressively measurable process $(X(t))_{t\in[0,T]}$ is called a mild solution to* (3.1) *if for every $t \in [0,T]$,*

$$\int_0^t \mathbb{E}(|S(t-s)b(s,X(s))| + \|S(t-s)\sigma(s,X(s))\|_{HS}^2)\mathrm{d}s < \infty, \tag{3.2}$$

and almost surely

$$X(t) = S(t)X(0) + \int_0^t S(t-s)b(s,X(s))\mathrm{d}s + \int_0^t S(t-s)\sigma(s,X(s))\mathrm{d}W(s).$$

F.-Y. Wang, *Harnack Inequalities for Stochastic Partial Differential Equations*, SpringerBriefs in Mathematics, DOI 10.1007/978-1-4614-7934-5_3,

To ensure the existence and uniqueness of the solution, we shall assume the following:

(A3.1) For every $s>0$ and $t\in[0,T]$, $S(s)b(t,0)\in\mathbb{H}$ with

$$\int_0^T \sup_{r\in[0,T]}|S(s)b(r,0)|^2\mathrm{d}s<\infty,$$

and there exists a positive function $K_b\in C((0,T])$ with

$$\phi_b(t):=\int_0^t K_b(s)\mathrm{d}s<\infty$$

such that

$$|S(t)(b(s,x)-b(s,y))|^2\le K_b(t)|x-y|^2,\quad s,t\in[0,T],x,y\in\mathbb{H}.$$

(A3.2) $\int_0^T \sup_{r\in[0,T]}\|S(s)\sigma(r,0)\|_{HS}^2\mathrm{d}s<\infty$, and there exists a positive function $K_\sigma\in C((0,T])$ with

$$\phi_\sigma(t):=\int_0^t K_\sigma(s)\mathrm{d}s<\infty$$

such that

$$\|S(t)(\sigma(s,x)-\sigma(s,y))\|_{HS}^2\le K_\sigma(t)|x-y|^2,\quad s,t\in[0,T],x,y\in\mathbb{H}.$$

Theorem 3.1.1 *Assume* **(A3.1)** *and* **(A3.2)**.

(1) *For every $X(0)\in L^2(\Omega\to\mathbb{H},\mathscr{F}_0,\mathbb{P})$,* (3.1) *has a unique mild solution $X(t)$, and there exists a constant $t_0\in(0,T]$ such that for every $n\ge1$,*

$$\sup_{t\in[0,T\wedge(nt_0)]}\mathbb{E}|X(t)|^2\le 6^n\mathbb{E}|X(0)|^2 \tag{3.3}$$
$$+12\Big(\sum_{i=1}^n 6^{n-i}\Big)\int_0^{t_0}\Big\{t_0\sup_{r\in[0,T]}|S(s)b(r,0)|^2+\sup_{r\in[0,T]}\|S(s)\sigma(r,0)\|_{HS}^2\Big\}\mathrm{d}s.$$

(2) *If there exists a constant $\varepsilon>0$ such that $\mathbb{E}|X(0)|^{2(1+\varepsilon)}<\infty$ and*

$$\int_0^T\Big\{K_\sigma(s)+K_b(s)+\sup_{r\in[0,T]}(\|S(s)\sigma(r,0)\|_{HS}^2+|S(s)b(r,0)|^2)\Big\}^{1+\varepsilon}\mathrm{d}s<\infty, \tag{3.4}$$

then

$$\mathbb{E}\sup_{t\in[0,T]}|X(t)|^{2(1+\varepsilon)}\le C(1+\mathbb{E}|X(0)|^{2(1+\varepsilon)}). \tag{3.5}$$

If, moreover,

$$\int_0^T s^{-\alpha}\Big\{K_\sigma(s)+\sup_{r\in[0,T]}\|S(s)\sigma(r,0)\|_{HS}^2\Big\}\mathrm{d}s<\infty \tag{3.6}$$

holds for some constant $\alpha \in (0,1)$ and all $r \in [0,T]$, $x \in \mathbb{H}$, then the solution has a continuous version.

Proof. (a) We first prove (3.3) for a mild solution $X(t)$ to (3.1). Obviously,

$$\mathbb{E}|X(t)|^2 \leq 3\mathbb{E}|X(0)|^2 + 3t\int_0^t \mathbb{E}|S(t-s)b(s,X(s))|^2 \mathrm{d}s \\ +3\int_0^T \mathbb{E}\|S(t-s)\sigma(s,X(s))\|_{HS}^2 \mathrm{d}s, \ t \in [0,T]. \tag{3.7}$$

Since by **(A3.1)** and **(A3.2)**,

$$|S(s)b(r,u)|^2 \leq 2|S(s)b(r,0)|^2 + 2K_b(s)|u|^2,$$
$$\|S(s)\sigma(r,u)\|_{HS}^2 \leq 2\|S(s)\sigma(r,0)\|_{HS}^2 + 2K_\sigma(s)|u|^2, \ u \in \mathbb{H},$$

it follows from (3.7) that

$$\sup_{s\in[0,t]} \mathbb{E}|X(s)|^2 \leq 3\mathbb{E}|X(0)|^2 + 6\{t\phi_b(t)+\phi_\sigma(t)\} \sup_{s\in[0,t]} \mathbb{E}|X(s)|^2 \\ +6\int_0^t \Big\{t \sup_{r\in[0,T]} |S(s)b(r,0)|^2 + \sup_{r\in[0,T]} \|S(s)\sigma(r,0)\|_{HS}^2\Big\}\mathrm{d}s.$$

Taking $t_0 \in (0,T]$ such that $6\{t_0\phi_b(t_0)+\phi_\sigma(t_0)\} \leq \frac{1}{2}$, we obtain

$$\sup_{s\in[0,t_0]} \mathbb{E}|X(s)|^2 \\ \leq 6\mathbb{E}|X(0)|^2 + 12\int_0^{t_0} \Big\{t_0 \sup_{r\in[0,T]} |S(s)b(r,0)|^2 + \sup_{r\in[0,T]} \|S(s)\sigma(r,0)\|_{HS}^2\Big\}\mathrm{d}s.$$

Therefore, letting $h(n) = \sup_{s\in[0,T\wedge(nt_0)]} \mathbb{E}|X(s)|^2$ and repeating the argument for the equation starting from time $T\wedge((n-1)t_0)$, we obtain

$$h(n) \leq 6h(n-1) + 12\int_0^{t_0} \Big\{t_0 \sup_{r\in[0,T]} |S(s)b(r,0)|^2 + \sup_{r\in[0,T]} \|S(s)\sigma(r,0)\|_{HS}^2\Big\}\mathrm{d}s, \ n \geq 1.$$

This implies (3.3).

(b) Existence and uniqueness. Let $\mathscr{J}$ be the set of $\mathbb{H}$-valued progressively measurable processes $\{Z(t)\}_{t\in[0,T]}$ such that $\sup_{t\in[0,T]} \mathbb{E}|Z(t)|^2 < \infty$. Then by **(A3.1)** and **(A3.2)**, for every $Z \in \mathscr{J}$,

$$J(Z)(t) := S(t)Z(0) + \int_0^t S(t-s)b(s,Z(s))\mathrm{d}s + \int_0^t S(t-s)\sigma(s,Z(s))\mathrm{d}W(s)$$

gives rise to an element in $\mathscr{J}$. By the fixed-point theorem and the local inversion theorem (see, e.g., [11, Proof of Theorem 7.4]), it remains to find $t_0 \in (0,T]$ such that

$$\sup_{t\in[0,t_0]} \mathbb{E}|J(Z)(t) - J(Y)(t)|^2 \leq \frac{1}{2} \sup_{t\in[0,t_0]} \mathbb{E}|Z(t)-Y(t)|^2, \ Z,Y \in \mathscr{J}. \tag{3.8}$$

In fact, by **(A3.1)** and **(A3.2)**, we have

$$\sup_{t\in[0,s]} \mathbb{E}|J(Z)(t)-J(Y(t))|^2 \le 2\{s\phi_b(s)+\phi_\sigma(s)\} \sup_{t\in[0,s]} \mathbb{E}|Z(t)-Y(t)|^2.$$

Since $\lim_{s\to 0}\{s\phi_b(s)+\phi_\sigma(s)\}=0$, there exists $t_0\in(0,T]$ such that (3.8) holds.

(c) Repeating the proof in (a) for $|X(t)|^{2(1+\varepsilon)}$ in place of $|X(t)|^2$ and using (3.4), we easily get (3.5). According to [11, Proposition 7.9], this and condition (3.6) imply that the solution has a continuous version. □

For $x\in\mathbb{H}$, let $X^x(t)$ be the unique mild solution to (3.1) for $X(0)=x$. We aim to investigate Harnack inequalities for the associated semigroup $(P_t)_{t\in[0,T]}$:

$$P_tf(x)=\mathbb{E}f(X^x(t)),\ \ x\in\mathbb{H},\ f\in\mathscr{B}_b(\mathbb{H}).$$

To this end, we will make use of finite-dimensional approximations, for which we shall need the following assumption:

(A3.3) $(A,\mathscr{D}(A))$ has a discrete spectrum, so that there exists an orthonormal basis $\{e_n, n\ge 1\}\subset\mathscr{D}(A)$ of $\mathbb{H}$ such that $-Ae_n=\lambda_n e_n, n\ge 1$, where $\lambda_n\ge 0, n\ge 0$ are all eigenvalues of $-A$ including multiplicities.

For $n\ge 1$, let $\mathscr{P}_n$ be the projection operator from $\mathbb{H}$ into $\mathbb{H}_n:=\operatorname{span}\{e_1,e_2,\dots,e_n\}$. Note that $\mathscr{P}_n$ commutes with the semigroup $S(t), t\ge 0$. Let

$$A_n=A|_{\mathbb{H}_n},\ \ b_n=\mathscr{P}_nb,\quad \sigma_n=\mathscr{P}_n\sigma. \tag{3.9}$$

Consider the following system of stochastic differential equations in $\mathbb{H}_n$:

$$\begin{cases} \mathrm{d}X^{(n)}(t) = A_nX^{(n)}(t)\mathrm{d}t+b_n(t,X^{(n)}(t))\mathrm{d}t+\sigma_n(t,X^{(n)}(t))\mathrm{d}W(t),\\ X^{(n)}(0) \ = \mathscr{P}_nX(0). \end{cases} \tag{3.10}$$

Note that

$$\sigma_n(t,X^{(n)}(t))\mathrm{d}W(t)=\sum_{i=1}^n\Big\{\sum_{j=1}^\infty\langle\sigma(t,X^{(n)}(t))e_j,e_i\rangle\mathrm{d}\langle W(t),e_j\rangle\Big\}e_i.$$

By **(A3.2)**, we may find a constant $C_n>0$ such that

$$\begin{aligned}&\sum_{i=1}^n\sum_{j=1}^\infty\langle(\sigma(t,x)-\sigma(t,y))e_j,e_i\rangle^2\\ &\le \mathrm{e}^{2\lambda_nT}\|S(T)(\sigma(t,x)-\sigma(t,y))\|_{HS}^2\le C_b(1+|x-y|^2),\\ &\sum_{i=1}^n\sum_{j=1}^\infty\langle\sigma(t,x)e_j,e_i\rangle^2\le \mathrm{e}^{2\lambda_nT}\|S(T)\sigma(t,x)\|_{HS}^2\le C_b(1+|x|^2),\ x,y\in\mathbb{H}_n, t\in[0,T].\end{aligned}$$

Similarly, **(A3.1)** implies that $b_n(t,\cdot)$ is Lipschitz continuous uniformly in $t\in[0,T]$. Therefore, it is well known that for any initial data, (3.10) has a unique strong solution.

Theorem 3.1.2 *Assume* **(A3.1)**, **(A3.2)**, *and* **(A3.3)**. *If* $\mathbb{E}|X(0)|^2 < \infty$, *then*

$$\lim_{n\to\infty} \mathbb{E}|X^{(n)}(t) - X(t)|^2 = 0, \quad t \in [0,T]. \tag{3.11}$$

Proof. Let $S_n(t)$ denote the semigroup generated by A_n. We have

$$S_n(t)x = \sum_{i=1}^{n} \mathrm{e}^{-\lambda_i t} \langle x, e_i\rangle e_i, \quad x \in \mathbb{H}_n,$$

and

$$X^{(n)}(t) = S_n(t)X^{(n)}(0) + \int_0^t S_n(t-s) b_n(s, X^{(n)}(s))\mathrm{d}s \\ + \int_0^t S_n(t-s)\sigma_n(s, X^{(n)}(s))\mathrm{d}W(s).$$

Subtracting X from $X^{(n)}$, we get

$$\mathbb{E}|X^{(n)}(t) - X(t)|^2 \le 3\mathbb{E}|S_n(t)X^{(n)}(0) - S(t)X(0)|^2 \\ +3t\mathbb{E}\int_0^t |S_n(t-s)b_n(s,X^{(n)}(s)) - S(t-s)b(s,X(s))|^2 \mathrm{d}s \\ +3\mathbb{E}\int_0^t \|S_n(t-s)\sigma_n(s,X^{(n)}(s)) - S(t-s)\sigma(s,X(s))\|_{HS}^2 \mathrm{d}s. \tag{3.12}$$

Now, since $S_n(t)X^{(n)}(0) = \mathscr{P}_n S(t)X(0)$ and $S_n(t-s)b_n = \mathscr{P}_n S(t-s)b$, we have

$$\mathbb{E}|S_n(t)X^{(n)}(0) - S(t)X(0)|^2 \\ = \mathbb{E}\sum_{k=n+1}^{\infty} \langle X(0), e_k\rangle^2 \mathrm{e}^{-2\lambda_k t} \le \mathbb{E}\sum_{k=n+1}^{\infty} \langle X(0), e_k\rangle^2, \tag{3.13}$$

and

$$\mathbb{E}\int_0^t |S_n(t-s)b_n(s,X^{(n)}(s)) - S(t-s)b(s,X(s))|^2 \mathrm{d}s \\ \le 2\mathbb{E}[\int_0^t |S_n(t-s)b_n(s,X^{(n)}(s)) - S_n(t-s)b_n(s,X(s))|^2 \mathrm{d}s \\ +2\mathbb{E}\int_0^t |S_n(t-s)b_n(s,X(s)) - S(t-s)b(s,X(s))|^2 \mathrm{d}s \tag{3.14} \\ \le 2\mathbb{E}\int_0^t |S(t-s)b(s,X^{(n)}(s)) - S(t-s)b(s,X(s))|^2 \mathrm{d}s \\ +2\mathbb{E}\int_0^T \sum_{k=n+1}^{\infty} \mathrm{e}^{-2\lambda_k(t-s)} \langle b(s,X(s)), e_k\rangle^2 \mathrm{d}s \\ \le 2\mathbb{E}\int_0^t K_b(t-s)|X^{(n)}(s) - X(s)|^2 \mathrm{d}s + 2\mathbb{E}\int_0^T \sum_{k=n+1}^{\infty} \mathrm{e}^{-2\lambda_k(t-s)} \langle b(s,X(s)), e_k\rangle^2 \mathrm{d}s.$$

To estimate the last term in (3.12), we observe that

$$\begin{aligned}
&\mathbb{E}\int_0^t \|S_n(t-s)\sigma_n(s,X^{(n)}(s)) - S_n(t-s)\sigma_n(s,X(s))\|_{HS}^2 \mathrm{d}s \\
&\leq \mathbb{E}\int_0^t \|S(t-s)\sigma(s,X^{(n)}(s)) - S(t-s)\sigma(s,X(s))\|_{HS}^2 \mathrm{d}s \\
&\leq \mathbb{E}\int_0^t K_\sigma(t-s)|X^{(n)}(s) - X(s)|^2 \mathrm{d}s,
\end{aligned} \tag{3.15}$$

where condition **(A3.2)** was used. Moreover,

$$\begin{aligned}
&\mathbb{E}\int_0^t \|S_n(t-s)\sigma_n(s,X(s)) - S(t-s)\sigma(s,X(s))\|_{HS}^2 \mathrm{d}s \\
&= \mathbb{E}\int_0^t \left\|(\mathscr{P}_n - \mathrm{Id})S(t-s)\sigma(s,X(s))\right\|_{HS}^2 \mathrm{d}s \\
&= \mathbb{E}\int_0^t \sum_{m=1}^\infty \sum_{k=n+1}^\infty \mathrm{e}^{-2\lambda_k(t-s)} \langle \sigma(s,X(s))e_m, e_k\rangle^2 \mathrm{d}s.
\end{aligned} \tag{3.16}$$

It follows from (3.15) and (3.16) that

$$\begin{aligned}
&\mathbb{E}\int_0^t \|S_n(t-s)\sigma_n(s,X^{(n)}(s)) - S(t-s)\sigma(s,X(s))\|_{HS}^2 \mathrm{d}s \\
&\leq 2\mathbb{E}\int_0^t K_\sigma(t-s)|X^{(n)}(s) - X(s)|^2 \mathrm{d}s \\
&\quad + 2\mathbb{E}\int_0^t \sum_{m=1}^\infty \sum_{k=n+1}^\infty \mathrm{e}^{-2\lambda_k(t-s)} \langle \sigma(s,X(s))e_m, e_k\rangle^2 \mathrm{d}s.
\end{aligned} \tag{3.17}$$

Putting (3.12), (3.13), (3.14), (3.17) together, we arrive at

$$\begin{aligned}
&\mathbb{E}|X^{(n)}(t) - X(t)|^2 \\
&\leq C\{a_n + c_n(t), d_n(t)\} + C\int_0^t (K_\sigma(t-s) + K_b(t-s))\mathbb{E}|X^{(n)}(s) - X(s)|^2 \mathrm{d}s
\end{aligned} \tag{3.18}$$

for some constant $C > 0$ and

$$\begin{aligned}
a_n &:= \mathbb{E}\sum_{k=n+1}^\infty \langle X_0, e_k\rangle^2, \\
c_n(t) &:= \mathbb{E}\int_0^t \sum_{k=n+1}^\infty \mathrm{e}^{-2\lambda_k(t-s)} \langle b(s,X(s)), e_k\rangle^2 \mathrm{d}s, \\
d_n(t) &:= \mathbb{E}\int_0^t \sum_{m=1}^\infty \sum_{k=n+1}^\infty \mathrm{e}^{-2\lambda_k(t-s)} \langle \sigma(s,X(s))e_m, e_k\rangle^2 \mathrm{d}s.
\end{aligned}$$

By (3.2) and the dominated convergence theorem, we see that $a_n + c_n(t), d_n(t) \to 0$ as $n \to \infty$. On the other hand, by Theorem 3.1.1, for every $T > 0$,

$$\sup_{n\geq 1}\sup_{0\leq t\leq T}\mathbb{E}|X^{(n)}(t)|^2+\sup_{0\leq t\leq T}\mathbb{E}|X(t)|^2<\infty. \tag{3.19}$$

So the function

$$g(t):=\limsup_{n\to\infty}\mathbb{E}|X^{(n)}(t)-X(t)|^2,\ t\geq 0,$$

is locally bounded. We will complete the proof of the theorem by showing that $g(t)=0$. Taking lim sup in (3.18) we obtain

$$g(t)\leq C\int_0^t (K_\sigma(t-s)+K_b(t-s))g(s)\mathrm{d}s. \tag{3.20}$$

Given any $\beta>0$. Multiplying (3.20) by $\mathrm{e}^{-\beta t}$ and integrating from 0 to T, we get

$$\begin{aligned}\int_0^T g(t)\mathrm{e}^{-\beta t}\mathrm{d}t &\leq C\int_0^T \mathrm{d}t\mathrm{e}^{-\beta t}\int_0^t (K_\sigma(t-s)+K_b(t-s))g(s)\mathrm{d}s\\ &= C\int_0^T \mathrm{e}^{-\beta s}g(s)\mathrm{d}s\int_s^T (K_\sigma(t-s)+K_b(t-s))\mathrm{e}^{-\beta(t-s)}\mathrm{d}t\\ &= C\int_0^T \mathrm{e}^{-\beta s}g(s)\mathrm{d}s\int_0^{T-s}(K_\sigma(u)+K_b(u))\mathrm{e}^{-\beta u}\mathrm{d}u\\ &\leq \left(C\int_0^T (K_\sigma(u)+K_b(u))\mathrm{e}^{-\beta u}\mathrm{d}u\right)\int_0^T \mathrm{e}^{-\beta s}g(s)\mathrm{d}s.\end{aligned} \tag{3.21}$$

Choosing $\beta>0$ sufficiently large that $C\int_0^T(K_\sigma(u)+K_b(u))\mathrm{e}^{-\beta u}\mathrm{d}u<1$, we deduce from (3.21) that $\int_0^T g(t)\mathrm{e}^{-\beta t}\mathrm{d}t=0$ and hence $g(t)=0$, a.e. By virtue of (3.20), we further conclude that $g(t)=0$ for every $t\in[0,T]$, and thus finish the proof. □

3.2 Additive Noise

In this section we consider the following semilinear SPDE with additive noise:

$$\mathrm{d}X(t)=\{AX(t)+b(t,X(t))\}\mathrm{d}t+\sigma(t)\mathrm{d}W(t), \tag{3.22}$$

where A, b, and σ satisfy **(A3.1)**, **(A3.2)**, and **(A3.3)**.

3.2.1 Harnack Inequalities and Bismut Formula

We first consider the Harnack inequalities for which we need the following ellipticity condition: $\sigma\sigma^*$ is invertible such that

$$\|\hat\sigma(t)\|^2\leq\lambda,\ \hat\sigma(t):=\sigma^*(t)(\sigma\sigma^*)^{-1}(t),\ t\geq 0, \tag{3.23}$$

holds for some constant $\lambda > 0$. When $\sigma(t)$ is invertible, this is equivalent to $\|\sigma^{-1}(t)\|^2 \le \frac{1}{\lambda}$. In general, there exist examples such that $\sigma\sigma^*$ is invertible but σ is not. For instance, let $\{e_i\}_{i\ge 1}$ be an orthonormal basis on $\mathbb{H}$ and let $\sigma e_1 = 0$ and $\sigma e_i = e_{i-1}, i \ge 2$, so that σ is not invertible. But $\sigma^* e_i = e_{i+1}$ for $i \ge 1$, and thus $\sigma\sigma^* = I$ is invertible.

Theorem 3.2.1 *Assume* **(A3.1)**, **(A3.2)**, **(A3.3)**, *and* (3.23). *If there exists a constant* $K \in \mathbb{R}$ *such that*

$$\langle (b(s,x) - b(s,y)), x - y\rangle \le K|x-y|^2 \tag{3.24}$$

holds for all $s \in [0,T]$ *and* $x,y \in \mathbb{H}$, *then for every strictly positive* $f \in \mathscr{B}_b(\mathbb{H})$ *and* $x,y \in \mathbb{H}$,

$$P_T \log f(y) \le \log P_T f(x) + \frac{K|x-y|^2}{\lambda(1-\mathrm{e}^{-2KT})}, \tag{3.25}$$

and for every $p > 1$,

$$(P_T f(y))^p \le (P_T f^p(x)) \exp\left[\frac{pK|x-y|^2}{\lambda(p-1)(1-\mathrm{e}^{-2KT})}\right]. \tag{3.26}$$

Consequently,

$$|\nabla P_T f| \le \delta\{P_T(f\log f) - (P_T f)\log P_T f\} + \frac{KP_T f}{\lambda\delta(1-\mathrm{e}^{-2KT})}, \quad \delta > 0. \tag{3.27}$$

Proof. By an approximation argument, it suffices to prove the result for $f \in C_b(\mathbb{H})$ with $\inf f > 0$. By Theorem 3.1.2, we have only to prove the assertions for $P_T^{(n)}$ associated to (3.10) in place of P_T. Let $T > 0$ and $x,y \in \mathbb{H}_n$ be fixed. For simplicity, we shall use $X(t)$ to denote $X^{(n)}(t)$, i.e., $X(t)$ solves (3.10) for $X(0) = x$, and let $Y(t)$ solve the equation

$$\mathrm{d}Y(t) = \left\{A_n Y(t) + b_n(t,Y(t)) + \eta(t)1_{[0,\tau)}(t)\frac{X(t)-Y(t)}{|X(t)-Y(t)|}\right\}\mathrm{d}t + \sigma_n(t)\mathrm{d}W(t)$$

for $Y(0) = y$, where $\eta \in C([0,\infty))$ is to be determined and $\tau := \inf\{t \ge 0 : X(t) = Y(t)\}$ is the coupling time. By (3.24), which holds also for b_n on $\mathbb{H}_n$, we obtain

$$\mathrm{d}|X(t)-Y(t)| \le K|X(t)-Y(t)|\mathrm{d}t - \eta(t)\mathrm{d}t, \quad t \in [0,\tau)\cap[0,T].$$

This implies that $\tau \le T$, and hence $X(T) = Y(T)$, provided that

$$\int_0^T \eta(t)\mathrm{e}^{-Kt} \ge |x-y|. \tag{3.28}$$

Now let $\mathrm{d}\mathbb{Q} = R\mathrm{d}\mathbb{P}$, where

$$R := \exp\left[-\int_0^\tau \langle \psi(t), \mathrm{d}W(t)\rangle - \frac{1}{2}\int_0^\tau |\psi(t)|^2\mathrm{d}t\right],$$

$$\psi(t) := \eta(t) 1_{[0,\tau)}(t) \frac{\hat{\sigma}(t)(X(t) - Y(t))}{|X(t) - Y(t)|}.$$

Since $\mathscr{P}_n(X(t) - Y(t)) = X(t) - Y(t)$, we may reformulate the equation for $Y(t)$ as

$$\mathrm{d}Y(t) = \{A_n Y(t) + b_n(t, Y(t))\}\mathrm{d}t + \sigma_n(t)\mathrm{d}\tilde{W}(t), \ Y(0) = y,$$

using the $\mathbb{Q}$-Brownian motion

$$\tilde{W}(t) := W(t) + \int_0^t \psi(s)\mathrm{d}s, \ t \geq 0.$$

Then (X, Y) is a coupling by change of measure with changed probability $\mathbb{Q}$ such that $X(T) = Y(T)$. Taking

$$\eta(t) = \frac{2K\mathrm{e}^{-Kt}|x - y|}{1 - \mathrm{e}^{-2KT}}, \ t \geq 1,$$

we see that (3.28) holds, and due to (3.23),

$$\mathbb{E}\{R \log R\} = \frac{1}{2}\mathbb{E}_{\mathbb{Q}} \int_0^\tau |\psi(t)|^2 \mathrm{d}t \leq \frac{1}{2\lambda} \int_0^T \eta(t)^2 \mathrm{d}t = \frac{K|x - y|^2}{\lambda(1 - \mathrm{e}^{-2KT})}.$$

Therefore, (3.25) follows from Theorem 1.1.1.

Moreover, we have

$$\begin{aligned}\left(\mathbb{E}R^{\frac{p}{p-1}}\right)^{p-1} &\leq \exp\left[\frac{p}{2\lambda(p-1)} \int_0^T \eta(t)^2 \mathrm{d}t\right] \left(\mathbb{E}\mathrm{e}^{-\frac{p}{p-1}\int_0^\tau \langle \Psi(t), \mathrm{d}W(t)\rangle - \frac{p^2}{2(p-1)^2}|\psi(t)|^2 \mathrm{d}t}\right)^{p-1} \\ &= \exp\left[\frac{pK|x - y|^2}{\lambda(p - 1)(1 - \mathrm{e}^{-2KT})}\right],\end{aligned}$$

which implies (3.26) according to Theorem 1.1.1.

Finally, applying Proposition 1.3.1 with $\delta_0 = 0$, $\beta(r, \cdot) = \frac{K}{\lambda r(1-\mathrm{e}^{-2KT})}$, and $\rho(x, y) = |x - y|$, we obtain (3.27) from (3.26). □

Remark 3.2.1. *Since the Girsanov transform used in the proof of Theorem* 3.2.1 *does not depend on the dimension n, with a finite-dimensional approximation we may simply assume that* $W(t)$ *(or equivalently* $\mathbb{H}$*) is finite-dimensional in the sequel.*

Next, we consider the Bismut formula.

Theorem 3.2.2 *Assume* **(A3.1)**, **(A3.2)**, *and*

$$\int_0^T \left\{\|\hat{\sigma}(t)\nabla_{S(t)z} b(t, \cdot)\|_\infty^2 + |\hat{\sigma}(t)S(t)z|^2\right\}\mathrm{d}t < \infty, \ z \in \mathbb{H}. \tag{3.29}$$

Let $\hat{\sigma}(t) = \sigma^*(t)(\sigma\sigma^*)^{-1}(t)$. *Then for every* $z, x \in \mathbb{H}$ *and* $f \in \mathscr{B}_b(\mathbb{H})$,

$$\nabla_z P_T f = \frac{1}{T}\mathbb{E}\left\{ f(X(T)) \int_0^T \left\langle \hat{\sigma}(t)\big(S(t)z + (T-t)\nabla_{S(t)z} b(t,\cdot)(X(t))\big),\, dW(t) \right\rangle \right\}.$$

Consequently, for every $\delta > 0$ *and positive* $f \in \mathscr{B}_b(\mathbb{H})$,

$$\begin{aligned} |\nabla_z P_T f| \le\ & \delta\{P_T(f\log f) - (P_T f)\log P_T f\} \\ & + \frac{P_T f}{2T^2\delta}\int_0^T \big(|\hat{\sigma}(t)S(t)z| + (T-t)\|\hat{\sigma}(t)\nabla_{S(t)z} b(t,\cdot)\|_\infty\big)^2 dt. \end{aligned}$$

Proof. The second assertion follows from the first and Young's inequality. Let $X(t)$ solve (3.22) for $X(0) = x$, and for $\varepsilon > 0$, let $X^\varepsilon(t)$ solve the equation

$$dX^\varepsilon(t) = \left\{ AX^\varepsilon(t) + b(t, X(t)) - \frac{\varepsilon}{T}S(t)z \right\} dt + \sigma(t)dW(t),\ \ X^\varepsilon(0) = x + \varepsilon z.$$

Then

$$X^\varepsilon(t) = X(t) + \frac{\varepsilon(T-t)}{T}S(t)z,\ \ t \in [0,T]. \tag{3.30}$$

In particular, $X^\varepsilon(T) = X(T)$. Set

$$\begin{aligned} \psi_\varepsilon(t) &= \hat{\sigma}(t)\Big(\frac{\varepsilon}{T}S(t)z + b(t, X^\varepsilon(t)) - b(t, X(t))\Big),\ \ t \in [0,T], \\ R_\varepsilon &= \exp\left[\int_0^T \langle \psi_\varepsilon(t), dW(t)\rangle - \frac{1}{2}\int_0^T |\psi_\varepsilon(t)|^2 dt\right]. \end{aligned}$$

Then

$$W^\varepsilon(t) := W(t) - \int_0^t \psi_\varepsilon(s)ds,\ \ t \in [0,T],$$

is a $d\mathbb{Q}_\varepsilon := R_\varepsilon d\mathbb{P}$-cylindrical Brownian motion, and $X^\varepsilon(t)$ solves

$$dX^\varepsilon(t) = \{AX^\varepsilon(t) + b(t, X^\varepsilon(t))\}dt + \sigma(t)dW^\varepsilon(t),\ \ X^\varepsilon(0) = x + \varepsilon z.$$

According to (3.30), **(A3.2)**, and (3.29), we see that

$$\lim_{\varepsilon \to 0} \frac{R_\varepsilon - 1}{\varepsilon} = \frac{1}{T}\int_0^T \left\langle \hat{\sigma}(t)\big(S(t)z + (T-t)\nabla_{S(t)z} b(t,\cdot)(X(t))\big),\, dW(t) \right\rangle$$

holds in $L^1(\mathbb{P})$. Therefore, the desired derivative formula follows from Theorem 1.1.2.

Finally, by the derivative formula and Young's inequality, and setting

$$M(t) = \frac{1}{\delta T}\int_0^t \left\langle \hat{\sigma}(t)\big(S(s)z + (T-s)\nabla_{S(s)z} b(s,\cdot)(X(s))\big), dW(s) \right\rangle,$$

we obtain

$$|\nabla_z P_T f| - \delta\{P_T(f\log f) - (P_T f)\log P_T f\} \le \delta(P_T f)\log \mathbb{E}e^{M(T)}$$
$$\le \delta(P_T f)\log \mathbb{E}\exp\left[M(T) - \frac{1}{2}\langle M\rangle(T)\right]$$
$$+\delta(P_T f)\log\exp\left[\frac{1}{2\delta^2 T^2}\int_0^T (|\hat\sigma(t)S(t)z| + (T-t)\|\hat\sigma(t)\nabla_{S(t)z}b(t,\cdot)\|_\infty)^2 dt\right]$$
$$= \frac{P_T f}{2T^2\delta}\int_0^T \left(|\hat\sigma(t)S(t)z| + (T-t)\|\hat\sigma(t)\nabla_{S(t)z}b(t,\cdot)\|_\infty\right)^2 dt.$$

□

To conclude this subsection, we present an example.

Example 3.2.3 (Stochastic reaction–diffusion equations) Let $(\Delta, \mathscr{D}(D))$ be the Dirichlet Laplacian on a bounded domain $D \subset \mathbb{R}^d$, and let $\alpha > \frac{d}{2}$ be a constant. Let $\mathbb{H} = L^2(\mathbf{m})$ for $\mathbf{m}$ be the normalized Lebesgue measure on D. Then $A := -(-\Delta)^\alpha$ satisfies **(A3.3)** with

$$\lambda_i \ge c i^{\frac{2\alpha}{d}}, \quad i \ge 1,$$

for some constant $c > 0$. Consider the equation

$$dX(t) = \{-(-\Delta)^\alpha X(t) + b(X(t))\}dt + \sigma dW(t),$$

where $W(t)$ is the cylindrical Brownian motion on $\mathbb{H}$, $\sigma e_i = \sigma_i e_i$, $i \ge 1$, for some constants $(\sigma_i)_{i\ge1}$ such that

$$\lambda := \inf_{i\ge1}\sigma_i^2 > 0, \quad \sum_{i=1}^\infty \frac{\sigma_i^2}{i^{\frac{2\alpha}{d}}} < \infty,$$

and

$$b(x) := \sum_{i=1}^\infty k_i \mathbf{m}(xe_i)e_i, \quad x \in \mathbb{H},$$

for some constants $(k_i)_{i\ge1}$ such that

$$K := \sup_{i\ge1} k_i < \infty, \quad \sum_{i=1}^\infty \frac{k_i^2}{i^{\frac{2\alpha}{d}}} < \infty.$$

Then (3.24) holds. Moreover, **(A3.1)** and **(A3.2)** hold for $K_\sigma = 0$ and

$$K_b(s) = \sum_{i=1}^\infty k_i^2 e^{-2\lambda_i s}, \quad \phi_b(t) = \sum_{i=1}^\infty \frac{k_i^2(1-e^{-2\lambda_i t})}{2\lambda_i} \le \sum_{i=1}^\infty \frac{k_i^2}{2\lambda_i} < \infty.$$

Therefore, the assertions in Theorem 3.2.1 hold.

Next, since $|\sigma(t)^{-1}S(t)z|^2 \le \frac{1}{\lambda}|z|^2$ and

$$\int_0^T |\nabla_{S(t)z}b|^2 dt = \int_0^T \left|\sum_{i=1}^\infty k_i e^{-\lambda_i t}\mathbf{m}(ze_i)e_i\right|^2 dt$$

$$= \sum_{i=1}^{\infty} \int_0^T k_i^2 e^{-2\lambda_i t} \mathbf{m}(ze_i)^2 dt \le \sum_{i=1}^{\infty} \frac{k_i^2}{2\lambda_i} \mathbf{m}(ze_i)^2 \le C|z|^2$$

holds for $C := \sup_{i\ge 1} \frac{k_i^2}{2\lambda_i} < \infty$, we see that (3.29) holds, so that Theorem 3.2.2 applies as well. In fact, for (3.29) we may weaken the condition $\lambda := \inf_{i\ge 1} \sigma_i^2 > 0$, since $\lambda' := \inf_{i\ge 1} \sigma_i^2 \lambda_i > 0$, because the latter implies

$$\int_0^T |\sigma(t)^{-1} S(t) z|^2 dt = \int_0^T \left| \sum_{i=1}^{\infty} \sigma_i^{-1} e^{-\lambda_i t} \mathbf{m}(ze_i) e_i \right|^2 dt$$
$$= \sum_{i=1}^{\infty} \int_0^T \sigma_i^{-2} e^{-2\lambda_i t} \mathbf{m}(ze_i)^2 dt \le \sum_{i=1}^{\infty} \frac{1}{2\sigma_i^2 \lambda_i} \mathbf{m}(ze_i)^2 \le \frac{|z|^2}{\lambda'} < \infty.$$

Observe that when $(k_i)_{i\ge 1}$ and $(\sigma_i)_{i\ge 1}$ are unbounded, b does not map $\mathbb{H}$ into $\mathbb{H}$, while σ is an unbounded linear operator on $\mathbb{H}$. See Sect. 3.3.2 below for a model with multiplicative noise.

3.2.2 Shift Harnack Inequalities and Integration by Parts Formula

The purpose of this section is to establish Driver's integration by parts formula and shift Harnack inequality for P_T associated to (3.22).

For $z \in \mathbb{H}$, let

$$z(t) = \int_0^t (e^{sA} z) ds, \quad t \ge 0.$$

Theorem 3.2.4 *Assume* **(A3.1)**, **(A3.2)**, **(A3.3)**, *and* (3.23). *If there exists a constant $K \ge 0$ such that*

$$|\nabla_z b(t,\cdot)(x)| \le K|z|, \quad x, z \in \mathbb{H}, t \in [0,T], \tag{3.31}$$

then:

(1) *For every $f \in C_b(\mathbb{H})$ having bounded directional derivatives,*

$$P_T(\nabla_{z(T)} f) = \mathbb{E}\left(f(X(T)) \int_0^T \left\langle \hat{\sigma}(t)\left(z - \nabla_{z(t)} b(t,\cdot)(X(t))\right), dW(t) \right\rangle \right).$$

Consequently, for every $\delta > 0$ and positive $f \in C_b(\mathbb{H})$ with bounded directional derivatives,

$$|P_T(\nabla_{z(T)} f)| \le \delta\{P_T(f \log f) - (P_T f) \log P_T f\} + \frac{|z|^2 P_T f}{2\delta\lambda}\left(T + KT^2 + \frac{K^2 T^3}{3}\right).$$

(2) *For every nonnegative $f \in \mathscr{B}_b(\mathbb{H})$ and $p > 1$,*

$$(P_T F)^p \le (P_T\{F(z(T) + \cdot)\}^p) \exp\left[\frac{p|z|^2}{2\lambda(p-1)}\left(T + T^2 K + \frac{T^3 K^2}{3}\right)\right].$$

(3) *For every positive* $f \in \mathscr{B}_b(\mathbb{H})$,

$$P_T \log f \le \log P_T\{f(z(T)+\cdot)\} + \frac{|z|^2}{2\lambda}\Big(T + T^2K + \frac{T^3K^2}{3}\Big).$$

Proof. For fixed $x \in \mathbb{H}$, let $X(t)$ solve (3.22) for $X(0) = x$. For $\varepsilon \in [0,1]$, let $X^\varepsilon(t)$ solve the equation

$$\mathrm{d}X^\varepsilon(t) = \Big\{AX^\varepsilon(t) + b(t,X(t)) + \varepsilon z\Big\}\mathrm{d}t + \sigma(t)\mathrm{d}W(t),\ \ t \ge 0, X^\varepsilon(0) = x.$$

Then it is easy to see that

$$X^\varepsilon(t) = X(t) + \varepsilon z(t),\ \ t \in [0,T]. \tag{3.32}$$

In particular, $X^\varepsilon(T) = X(T) + \varepsilon z(T)$. Next, let

$$\begin{aligned} R_\varepsilon = \exp\Bigg[&-\int_0^T \Big\langle \hat\sigma(t)\Big\{\varepsilon z + b(t,X(t)) - b(t,X^\varepsilon(t))\Big\}, \mathrm{d}W(t)\Big\rangle \\ &-\frac{1}{2}\int_0^T \Big|\hat\sigma(t)\Big\{\varepsilon z + b(t,X(t)) - b(t,X^\varepsilon(t))\Big\}\Big|^2 \mathrm{d}t\Bigg]. \end{aligned}$$

By Girsanov's theorem, under the weighted probability $\mathbb{Q}_\varepsilon := R_\varepsilon \mathbb{P}$, the process

$$W^\varepsilon(t) := W(t) + \int_0^t \hat\sigma(s)\Big(\varepsilon z + b(s,X_s) - b(s,X_s^\varepsilon)\Big)\mathrm{d}s,\ \ t \in [0,T],$$

is a cylindrical Brownian motion. So $(X(t), X^\varepsilon(t))$ is a coupling by change of measure with changed probability $\mathbb{Q}_\varepsilon$. Then the desired integration by parts formula follows from Theorem 1.1.3, since $R_0 = 1$ and due to (3.32),

$$\frac{\mathrm{d}}{\mathrm{d}\varepsilon}R^\varepsilon\Big|_{\varepsilon=0} = -\int_0^T \Big\langle \hat\sigma(t)\Big(z - \nabla_{z(t)}b(t,\cdot)(X(t))\Big), \mathrm{d}W(t)\Big\rangle$$

holds in $L^1(\mathbb{P})$. Then the inequality in (1) follows because combining this formula with (3.23) and (3.31), we obtain

$$\begin{aligned} &|P_T(\nabla_{z_T} f)| - \delta\{P_T(f\log f) - (P_T f)\log P_T f\} \\ &\le \delta(P_T f)\log \mathbb{E}\exp\Bigg[\frac{1}{\delta}\int_0^T \Big\langle \hat\sigma(t)\Big(z - \nabla_{z_t}b(t,\cdot)(X(t))\Big), \mathrm{d}W(t)\Big\rangle\Bigg] \\ &\le \delta(P_T f)\log \mathbb{E}e^{\frac{1}{\delta}\int_0^T \langle \hat\sigma(t)(z-\nabla_{z_t}b(t,\cdot)(X(t))), \mathrm{d}W(t)\rangle - \frac{1}{2\delta^2}\int_0^T |\hat\sigma(t)(z-\nabla_{z_t}b(t,\cdot)(X(t)))|^2\mathrm{d}t} \\ &\quad + \frac{P_T f}{2\lambda\delta}\int_0^T \big(|z| + K|z(t)|\big)^2\mathrm{d}t \\ &= \frac{P_T f}{2\lambda\delta}\int_0^T \big(|z| + K|z(t)|\big)^2\mathrm{d}t = \frac{|z|^2 P_T f}{2\lambda\delta}\Big(T + KT^2 + \frac{K^2T^3}{3}\Big). \end{aligned}$$

Next, by the inequality in (1), we may apply Proposition 1.3.2 with $r = 1$, $e = z(T)$, and

$$\beta_e(\delta,\cdot) = \frac{|z|^2}{2\lambda\delta}\Big(T + KT^2 + \frac{K^2T^3}{3}\Big)$$

to prove (2). Finally, (3) follows from (2) according to Theorem 1.3.5. □

3.3 Multiplicative Noise: The Log-Harnack Inequality

In this section we consider (3.1) for b, σ, and A satisfying assumptions **(A3.1)**, **(A3.2)**, and **(A3.3)**. Since $\phi_b(t) + \phi_\sigma(t) \to 0$ as $t \to 0$, we have

$$t_0 := \sup\Big\{t > 0 : t\phi_b(t) + \phi_\sigma(t) \le \frac{1}{6}\Big\} > 0.$$

3.3.1 The Main Result

The following is the main result of this section, where when $t_0 = \infty$, $t_0(6^{\frac{t}{t_0}} - 1)$ and $t_0(1 - 6^{-\frac{t}{t_0}})$ are understood as their limits by letting $t_0 \to \infty$, i.e.,

$$t_0(6^{\frac{t}{t_0}} - 1) = t_0(1 - 6^{-\frac{t}{t_0}}) = t\log 6.$$

Theorem 3.3.1 *Assume* **(A3.1)**, **(A3.2)**, *and* **(A3.3)**. *Then:*

(1) *For every $f \in C_b^1(\mathbb{H})$,*

$$|\nabla P_t f|^2 \le 6^{1+\frac{t}{t_0}} P_t |\nabla f|^2, \quad t \ge 0. \tag{3.33}$$

(2) *Let $b(t,x) = b(x), \sigma(t,x) = \sigma(x)$ be independent of t. If there exists a constant $\lambda > 0$ such that $\|(\sigma\sigma^*)^{-1}\| \le \frac{1}{\lambda}$, then for every strictly positive $f \in \mathscr{B}_b(\mathbb{H})$,*

$$P_t \log f(y) \le \log P_t f(x) + \frac{3\log 6}{\lambda t_0(1 - 6^{-\frac{t}{t_0}})}|x-y|^2, \quad x, y \in \mathbb{H}, t \in [0,T].$$

Consequently, for every $f \in \mathscr{B}_b(\mathbb{H})$,

$$|\nabla P_t f|^2 \le \frac{3\log 6}{t_0\lambda(1 - 6^{-\frac{t}{t_0}})}\{P_t f^2 - (P_t f)^2\}, \quad t \in [0,T].$$

(3) *If $b(t,x) = b(x)$ and $\sigma(t,x) = \sigma(x)$ are independent of t, and $\|\sigma(\cdot)^*\|^2 \le \bar{\lambda}$ holds for some constant $\bar{\lambda} > 0$, then*

$$P_t f^2 - (P_t f)^2 \le \frac{12\bar{\lambda} t_0 (6^{\frac{t}{t_0}} - 1)}{\log 6} P_t |\nabla f|^2, \ \ f \in C_b^1(\mathbb{H}), t \in [0,T].$$

Proof. According to Theorem 3.1.2, we need to prove the result only for $P_t^{(n)}$ associated to the finite-dimensional equation (3.10). Indeed, as explained in Remark 3.2.1, we may and do assume that $\mathbb{H}$ itself is finite-dimensional, i.e., $\mathbb{H} = \mathbb{R}^n$ for some $n \ge 1$. In this case, **(A3.1)** and **(A3.2)** imply that $b(t,\cdot)$ and $\sigma(t,\cdot)$ are Lipschitz continuous uniformly in $t \in [0,T]$. By **(A3.1)**, **(A3.2)**, and a standard approximation argument, we may and do assume that they are smooth in the second variable with

$$|S(s)\nabla_v b(t,\cdot)|^2 \le K_b(s)|v|^2, \ \|S(s)\nabla_v \sigma(t,\cdot)\|_{HS}^2 \le K_\sigma(s)|v|^2, \ \ s,t > 0, v \in \mathbb{R}^n. \tag{3.34}$$

In this case, the derivative process (recall that $X^x(t)$ is the solution to (3.1) for $X(0) = x$)

$$\nabla_v X(t) := \lim_{\varepsilon \downarrow 0} \frac{X^{\cdot + \varepsilon v}(t) - X(t)}{\varepsilon}$$

solves the equation

$$\mathrm{d}\nabla_v X(t) = \Big(A\nabla_v X(t) + \{\nabla_{\nabla_v X(t)} b(t,\cdot)\}(X(t))\Big)\mathrm{d}t + \{\nabla_{\nabla_v X(t)} \sigma(t,\cdot)\}(X(t))\mathrm{d}W_t$$

with $\nabla_v X(0) = v$. Since ∇b and $\nabla \sigma$ are bounded, this implies that

$$\sup_{s \in [0,t]} \mathbb{E}|\nabla_v X(s)|^2 < \infty, \ \ t \ge 0.$$

We aim to find an upper bound of $\mathbb{E}|\nabla_v X(t)|^2$ independent of the dimension n so that it can be passed to the infinite-dimensional setting. To this end, let us observe that for every $s_0 \ge 0$,

$$\begin{aligned} \nabla_v X(t) = \ & S(t-s_0)\nabla_v X(s_0) + \int_{s_0}^t S(t-s)\{\nabla_{\nabla_v X(s)} b(s,\cdot)\}(X(s))\mathrm{d}s \\ & + \int_{s_0}^t S(t-s)\{\nabla_{\nabla_v X(s)} \sigma(s,\cdot)\}(X(s))\mathrm{d}W_s, \ \ t \ge s_0. \end{aligned} \tag{3.35}$$

Combining this with (3.34), we obtain

$$\begin{aligned} \mathbb{E}|\nabla_v X(t)|^2 \ & \le 3\mathbb{E}|\nabla_v X(s_0)|^2 + 3\int_{s_0}^t \{(t-s_0)K_b(t-s) + K_\sigma(t-s)\}\mathbb{E}|\nabla_v X(s)|^2 \mathrm{d}s \\ & \le 3\mathbb{E}|\nabla_v X(s_0)|^2 + \{3(t-s_0)\phi_b(t-s_0) + 3\phi_\sigma(t-s_0)\} \sup_{s \in [s_0,t]} \mathbb{E}|\nabla_v X(s)|^2 \end{aligned}$$

for $t \ge s_0$. Since the resulting upper bound is increasing in $t \ge s_0$, it follows that

$$\begin{aligned} \sup_{s \in [s_0,t]} \mathbb{E}|\nabla_v X(s)|^2 \le \ & 3\mathbb{E}|\nabla_v X(s_0)|^2 + \{3(t-s_0)\phi_b(t-s_0) \\ & + 3\phi_\sigma(t-s_0)\} \sup_{s \in [s_0,t]} \mathbb{E}|\nabla_v X(s)|^2 \end{aligned}$$

holds for $t \geq s_0$. Taking $t = s_0 + t_0$ in this inequality leads to

$$\sup_{s \in [s_0, s_0+t_0]} \mathbb{E}|\nabla_v X(s)|^2 \leq 6\mathbb{E}|\nabla_v X(s_0)|^2, \quad s_0 \geq 0.$$

Therefore,

$$\mathbb{E}|\nabla_v X(t)|^2 \leq 6^{\frac{t+t_0}{t_0}} |v|^2, \quad t \geq 0, v \in \mathbb{R}^n. \tag{3.36}$$

With this estimate in hand, we are able to complete the proof as follows.

The gradient inequality (3.33) is implied by (3.36) and the Schwarz inequality:

$$|\nabla_v P_t f|^2 = \left|\nabla_v \mathbb{E} f(X(t))\right|^2 = \left|\mathbb{E}\langle \nabla f(X(t)), \nabla_v X(t)\rangle\right|^2 \leq 6^{\frac{t+t_0}{t_0}} |v|^2 P_t |\nabla f|^2.$$

Noting that $\|(\sigma\sigma^*)^{-1}\| \leq \frac{1}{\lambda}$ implies

$$\Gamma(f) = \frac{1}{2}|\sigma^* \nabla f|^2 \geq \frac{\lambda}{2}|\nabla f|^2,$$

when $b(t,x)$ and $\sigma(t,x)$ are independent of t, the first inequality in (2) follows from the gradient inequality (3.33) and Theorem 1.3.6 with

$$g(s) := \frac{1 - 6^{-\frac{s}{t_0}}}{1 - 6^{-\frac{t}{t_0}}},$$

while the second inequality then follows from Proposition 1.3.8. Finally, (3) follows from (3.33) by noting that

$$\frac{\mathrm{d}}{\mathrm{d}s} P_s (P_{t-s} f)^2 = 2 P_s |\sigma^* \nabla P_{t-s} f|^2 \leq 2\bar{\lambda} P_s |\nabla P_{t-s} f|^2 \leq 2\bar{\lambda} 6^{\frac{t-s-t_0}{t_0}} P_t |\nabla f|^2, \quad s \in [0,t].$$

□

3.3.2 Application to White-Noise-Driven SPDEs

In this subsection, we apply Theorem 3.3.1 to stochastic reaction–diffusion equations driven by space-time white noise; see [12] and references therein. When the noise is additive, Harnack inequalities with power have been derived in [74], where a reflection is included in the equation.

Consider the following stochastic reaction–diffusion equation on a bounded domain $D \subset \mathbb{R}^d (d \geq 1)$:

$$\begin{aligned} \frac{\partial X(t)(v)}{\partial t} &= -(-\Delta)^\alpha X(t)(v) + \psi(X(t)(v)) + \phi(X(t)(v)) \frac{\partial^{1+d}}{\partial t \partial v_1 \cdots \partial v_d} W(t,v), \\ X(0) &= g, \quad X(t)|_{\partial D} = 0, \quad v = (v_1, \cdots, v_d) \in D, \end{aligned} \tag{3.37}$$

where $\alpha > 0$ is a constant, $W(t,v)$ is a Brownian sheet on $\mathbb{R}^{d+1}$, Δ is the Dirichlet Laplacian on D, and ϕ, ψ are Lipschitz functions on $\mathbb{R}$, i.e., there exists a constant $c > 0$ such that

$$|\psi(r) - \psi(s)| \le c|r-s|, \quad |\phi(r) - \phi(s)| \le c|r-s|, \quad r,s \in \mathbb{R}. \tag{3.38}$$

To apply Theorem 3.3.1 to the present model, we reformulate the equation using cylindrical Brownian motion on $\mathbb{H} := L^2(\mathbf{m})$, where $\mathbf{m}$ is Lebesgue measure on D. Let $A = -(-\Delta)^\alpha$. Then $-A$ has a discrete spectrum with eigenvalues $\{\lambda_n\}_{n\ge1}$ satisfying

$$\frac{n^{\frac{2\alpha}{d}}}{C} \le \lambda_n \le C n^{\frac{2\alpha}{d}}, \quad n \ge 1, \tag{3.39}$$

for some constant $C > 1$. Let $\{e_n\}_{n\ge1}$ be the corresponding unit eigenfunctions. Since e_m is independent of α, letting $\alpha = 1$ and using the classical Dirichlet heat kernel bound, we obtain

$$\|e_m\|_\infty = \mathrm{e}\|S(\lambda_m^{-1})e_m\|_\infty \le \mathrm{e}\|S(\lambda_m^{-1})\|_{L^2(\mathbf{m})\to L^\infty(\mathbf{m})} \le c_1\lambda_m^{\frac{d}{4}} \le c_2\sqrt{m}, \quad m \ge 1, \tag{3.40}$$

for some constants $c_1, c_2 > 0$.

Now define a sequence of independent Brownian motions by

$$\beta_n(t) = \int_{[0,t]\times D} e_n(v) W(\mathrm{d}s, \mathrm{d}v), \quad n \ge 1.$$

Then

$$W(t) := \sum_{n=1}^\infty \beta_n(t) e_n$$

is a cylindrical Brownian motion on $\mathbb{H} := L^2(\mathbf{m})$. Let

$$b(u)(v) = \psi(u(v)), \quad \{\sigma(u)x\}(\xi) = \phi(u(v)) \cdot x(v), \quad u,x \in \mathbb{H}, v \in D.$$

It is easy to see that the reaction–diffusion equation (3.37) can be reformulated as

$$\mathrm{d}X(t) = AX(t)\mathrm{d}t + b(X(t))\mathrm{d}t + \sigma(X(t))\mathrm{d}W(t). \tag{3.41}$$

Obviously, σ takes values in the space of bounded linear operators on $\mathbb{H}$ if and only if ϕ is bounded.

Theorem 3.3.2 *Consider* (3.41) *with the above A, b, and σ such that $\phi^2 \ge \lambda$ for some constant $\lambda > 0$.*

(1) *If $\alpha > d$, then the assertions in Theorem* 3.3.1 *hold for some constant $t_0 > 0$.*
(2) *Let $D = \prod_{i=1}^d [a_i, b_i]$ for some $b_i > a_i$, $1 \le i \le d$. If $\alpha > \frac{d}{2}$, then the assertions in Theorem* 3.3.1 *hold for some constant $t_0 > 0$.*
(3) *In the situations of* (1) *and* (2), *there exists a constant $\varepsilon_0 > 0$ such that P_t has a unique invariant probability measure, provided that*

$$|\phi(s)|^2 + |\psi(s)|^2 \le \varepsilon_0 |s|^2 + C_0, \quad s \in \mathbb{R}, \tag{3.42}$$

holds for some constant $C_0 > 0$.

Proof. Since **(A3.3)** is obvious due to (3.39), and $|\sigma^* x|^2 \ge \lambda |x|^2$ follows from $\phi^2 \ge \lambda$, for (1) and (2) it suffices to verify **(A3.1)** and **(A3.2)** in the specific situations of D. By the contraction of $S(t)$ and (3.38), we have

$$|S(t)(b(x) - b(y))| \le c|x-y|, \quad x, y \in \mathbb{H},$$

where $|\cdot|$ is now the L^2-norm on D. Then **(A3.1)** holds for $K_b \equiv c^2$.

Below, we verify **(A3.2)** and the existence of the invariant probability measure claimed in (3).

(1) By the definition of σ, (3.38), and (3.40), we have

$$\begin{aligned} \|S(t)(\sigma(x) - \sigma(y))\|_{HS}^2 &= \sum_{n=1}^{\infty}\sum_{m=1}^{\infty} \langle S(t)(\sigma(x) - \sigma(y))e_n, e_m\rangle^2 \\ &= \sum_{m=1}^{\infty} \mathrm{e}^{-2t\lambda_m} \sum_{n=1}^{\infty} \langle (\sigma(x) - \sigma(y))e_n, e_m\rangle^2 = \sum_{m=1}^{\infty} \mathrm{e}^{-2t\lambda_m} |(\sigma(x) - \sigma(y))^* e_m|^2 \\ &= \sum_{m=1}^{\infty} \mathrm{e}^{-2t\lambda_m} \int_D |(\phi(x(v)) - \phi(y(v)))e_m(v)|^2 \mathrm{d}v \le c^2 |x-y|^2 \sum_{m=1}^{\infty} \|e_m\|_\infty^2 \mathrm{e}^{-\delta t m^{\frac{2\alpha}{d}}} \end{aligned} \tag{3.43}$$

for some constant $\delta > 0$. Combining this with (3.40), we obtain

$$\|S(t)(\sigma(x) - \sigma(y))\|_{HS}^2 \le C|x-y|^2 \sum_{m=1}^{\infty} m\mathrm{e}^{-\delta t m^{\frac{2\alpha}{d}}}$$

for some constant $C > 0$. Moreover,

$$\int_0^t \|S(s)\sigma(0)\|_{HS}^2 = \phi(0)^2 \int_0^t \|S(s)\|_{HS}^2 \mathrm{d}s \le C' \sum_{m=1}^{\infty} m^{-\frac{2\alpha}{d}} < \infty \tag{3.44}$$

holds for some constant $C' > 0$, provided that $2\alpha > d$. Therefore, if $\alpha > d$, then **(A3.2)** holds for

$$K_\sigma(t) = C \sum_{m=1}^{\infty} m\mathrm{e}^{-\delta t m^{\frac{2\alpha}{d}}} \tag{3.45}$$

for some constant $C > 0$, since in this case,

$$\int_0^\infty \sum_{m=1}^{\infty} m\mathrm{e}^{-\delta t m^{\frac{2\alpha}{d}}} \mathrm{d}t = \frac{1}{\delta} \sum_{m=1}^{\infty} \frac{1}{m^{\frac{2\alpha - d}{d}}} < \infty.$$

(2) When $D = \prod_{i=1}^d [a_i, b_i]$ for some $b_i > a_i$, $1 \le i \le d$, the eigenfunctions $\{e_m\}_{m \ge 1}$ are uniformly bounded, i.e., $\|e_m\|_\infty \le C$ holds for some constant $C > 0$ and all $m \ge 1$. Combining this with (3.43), we obtain

$$\|S(t)(\sigma(x)-\sigma(y))\|_{HS}^2 \le C|x-y|^2 \sum_{m=1}^{\infty} e^{-\delta t m^{\frac{2\alpha}{d}}}$$

for some constants $C, \delta > 0$. Combining this with (3.44), we conclude that **(A3.2)** holds for

$$K_\sigma(t) = C\sum_{m=1}^{\infty} e^{-\delta t m^{\frac{2\alpha}{d}}} \tag{3.46}$$

for some $C > 0$, provided $\alpha > \frac{d}{2}$.

(3) The uniqueness of the invariant probability measure follows from the log-Harnack inequality according to Theorem 1.4.1. So it suffices to prove the existence by verifying conditions (i)–(iv) in [12, Theorem 6.1.2]. By (3.38), (3.39), and **(A3.2)**, conditions (i) and (iii) hold. Moreover, (iv) is implied by (3.3). It remains to verify condition (ii), i.e.,

$$\int_0^1 s^{-\varepsilon} K_\sigma(s)\mathrm{d}s < \infty \text{ for some } \varepsilon \in (0,1). \tag{3.47}$$

Let K_σ be as in (3.45) with $\alpha > d$. Then for $\varepsilon \in (0, \frac{\alpha-d}{\alpha})$,

$$\begin{aligned}\int_0^1 s^{-\varepsilon} K_\sigma(s)\mathrm{d}s &= C\sum_{m=1}^{\infty} m\int_0^1 s^{-\varepsilon} \exp\big[-\delta s m^{\frac{2\alpha}{d}}\big]\mathrm{d}s\\ &\le C\sum_{m=1}^{\infty} m\bigg(\int_0^{m^{-\frac{2\alpha}{d}}} s^{-\varepsilon}\mathrm{d}s + m^{\frac{2\alpha\varepsilon}{d}}\int_{m^{-\frac{2\alpha}{d}}}^1 \exp\big[-\delta s m^{\frac{2\alpha}{d}}\big]\mathrm{d}s\bigg)\\ &\le C'\sum_{m=1}^{\infty} m^{1-\frac{2(1-\varepsilon)\alpha}{d}} < \infty,\end{aligned}$$

where $C' > 0$ is a constant. Similarly, (3.47) holds for K_σ in (3.46) with $\alpha > \frac{d}{2}$ and $\varepsilon \in (0, \frac{2\alpha-d}{2\alpha})$. □

3.4 Multiplicative Noise: Harnack Inequality with Power

In this section, besides **(A3.1)–(A3.3)**, we assume moreover

(A3.4) There exists a constant $K \in \mathbb{R}$ such that for every $x, y \in \mathbb{H}_\infty := \cup_{n\ge 1}\mathbb{H}_n$ and $t \in [0,T]$,

$$\|\sigma(t,x)-\sigma(t,y)\|_{HS}^2 + 2\langle A(x-y)+b(t,x)-b(t,y), x-y\rangle \le K|x-y|^2.$$

(A3.5) $\sigma\sigma^*$ is invertible such that $\|\hat{\sigma}(t,x)\|^2 \le \frac{1}{\lambda}$ holds for some constant $\lambda > 0$ and all $x \in \mathbb{H}$ and $t \in [0,T]$, where $\hat{\sigma} = \sigma^*(\sigma\sigma^*)^{-1}$.

Theorem 3.4.1 *Assume* **(A3.1)–(A3.5)**.

(1) *For every strictly positive* $f \in \mathscr{B}_b(\mathbb{H})$ *and* $x,y \in \mathbb{H}$,

$$P_T \log f(y) \le \log P_T f(x) + \frac{K|x-y|^2}{2\lambda(1-\mathrm{e}^{-KT})}.$$

(2) *If, moreover,*

$$\big|(\sigma(t,x)-\sigma(t,y))(x-y)\big| \le \delta|x-y|, \ \ x,y \in \mathbb{H}_\infty, \ t \in [0,T], \tag{3.48}$$

holds for some constant $\delta \ge 0$, *then for* $p > (1+\frac{\delta}{\sqrt{\lambda}})^2$ *and* $\delta_p := \max\{\delta, \frac{\sqrt{\lambda}}{2}(\sqrt{p}-1)\}$,

$$(P_T f(y))^p \le (P_T f^p(x)) \exp\left[\frac{K\sqrt{p}(\sqrt{p}-1)|x-y|^2}{4\delta_p[(\sqrt{p}-1)\sqrt{\lambda}-\delta_p](1-\mathrm{e}^{-KT})}\right]$$

holds for all $x,y \in \mathbb{H}$ *and positive* $f \in \mathscr{B}_b(\mathbb{H})$.

By Theorem 3.1.2, it suffices to prove the result for $\mathbb{H} = \mathbb{R}^d$ for some $d \ge 1$. In this case, **(A3.1)** and **(A3.2)** imply that $b(t,\cdot)$ and $\sigma(t,\cdot)$ are Lipschitz continuous uniformly in $t \in [0,T]$. Letting $\bar{b}(t,x) = Ax + b(t,x)$, we reduce (3.1) to the following SDE on $\mathbb{R}^d$:

$$\mathrm{d}X(t) = \sigma(t,X(t))\mathrm{d}W(t) + \bar{b}(t,X(t))\mathrm{d}t, \tag{3.49}$$

which has a unique strong solution $X^x(t)$ starting from any $x \in \mathbb{R}^d$. To prove this theorem, we first introduce the construction of coupling by change of measure, which is very technical in the multiplicative noise setting.

3.4.1 Construction of the Coupling

Let $x,y \in \mathbb{R}^d$, $T > 0$, and $p > (1+\frac{\delta}{\sqrt{\lambda}})^2$ be fixed such that $x \ne y$. For $\theta \in (0,2)$, let

$$\gamma_\theta(t) = \frac{2-\theta}{K}(1-\mathrm{e}^{K\,(t-T)}), \ \ t \in [0,T]. \tag{3.50}$$

Then γ_θ is smooth and strictly positive on $[0,T)$ such that

$$2 - K\,\gamma_\theta(t) + \gamma_\theta'(t) = \theta, \ \ t \in [0,T]. \tag{3.51}$$

Let $X(t)$ solve (3.49) for $X(0) = x$, and consider the equation

$$\begin{aligned}\mathrm{d}Y(t) = \ &\bar{b}(t,Y(t))\mathrm{d}t + \frac{1}{\gamma_\theta(t)}\sigma(t,Y(t))\hat{\sigma}(t,X(t))(X(t)-Y(t))\mathrm{d}t\\ &+\sigma(t,Y(t))\mathrm{d}W(t), \ \ Y(0) = y.\end{aligned} \tag{3.52}$$

Since for $t \in [0,T)$, the coefficients in (3.49) and (3.52) are locally Lipschitz in (x,y), the coupling $(X(t),Y(t))$ is a well-defined continuous process for $t \in [0,T)$. Let

$$\mathrm{d}\tilde{W}(t) = \mathrm{d}W(t) + \frac{1}{\gamma_\theta(t)}\hat{\sigma}(t,X(t))(X(t)-Y(t))\mathrm{d}t, \ t < T.$$

If

$$R(s) := \exp\bigg[-\int_0^s \gamma_\theta(t)^{-1}\langle\hat{\sigma}(t,X(t))(X(t)-Y(t)),\mathrm{d}W(t)\rangle \\ -\frac{1}{2}\int_0^s \gamma_\theta(t)^{-2}|\hat{\sigma}(t,X(t))(X(t)-Y(t))|^2\mathrm{d}t\bigg]$$

is a uniformly integrable martingale for $s \in [0,T)$, then by the martingale convergence theorem, $R(T) := \lim_{t\uparrow T} R(t)$ exists and $\{R(t)\}_{t\in[0,T]}$ is a martingale. In this case, by Girsanov's theorem, $\{\tilde{W}(t)\}_{t\in[0,T]}$ is a d-dimensional Brownian motion under the probability $\mathrm{d}\mathbb{Q} := R(T)\mathrm{d}\mathbb{P}$. Rewrite (3.49) and (3.52) as

$$\mathrm{d}X(t) = \sigma(t,X(t))\mathrm{d}\tilde{W}(t) + \bar{b}(t,X(t))\mathrm{d}t - \frac{X(t)-Y(t)}{\gamma_\theta(t)}\mathrm{d}t, \ \ X_0 = x, \tag{3.53}$$

$$\mathrm{d}Y(t) = \sigma(t,Y(t))\mathrm{d}\tilde{W}(t) + \bar{b}(t,Y(t))\mathrm{d}t, \ \ Y_0 = y. \tag{3.54}$$

Since $\int_0^T \gamma_\theta(t)^{-1}\mathrm{d}t = \infty$, we will see that the additional drift $-\frac{X(t)-Y(t)}{\gamma_\theta(t)}\mathrm{d}t$ is strong enough to force the coupling to be successful up to time T. Thus, below we first prove the uniform integrability of $\{R(s)\}_{s\in[0,T)}$ with respect to $\mathbb{P}$, so that $R(T) := \lim_{s\uparrow T} R(s)$ exists and $\mathrm{d}\mathbb{Q} := R(T)\mathrm{d}\mathbb{P}$ is a probability measure.

Lemma 3.4.2 *We have*

$$\sup_{s\in[0,T)} \mathbb{E}\{R(s)\log R(s)\} \le \frac{K\,|x-y|^2}{2\lambda\theta(2-\theta)(1-\mathrm{e}^{-K\,T})}.$$

Consequently, $R(T) := \lim_{s\uparrow T} R(s)$ *exists and* $\{R(s)\}_{s\in[0,T]}$ *is a uniformly integrable martingale.*

Proof. Let $s \in [0,T)$ be fixed. By (3.54), **(A3.4)**, and Itô's formula,

$$\mathrm{d}|X(t)-Y(t)|^2 \le 2\langle(\sigma(t,X(t))-\sigma(t,Y(t)))(X(t)-Y(t)),\mathrm{d}\tilde{W}(t)\rangle \\ +K\,|X(t)-Y(t)|^2\mathrm{d}t - \frac{2}{\gamma_\theta(t)}|X(t)-Y(t)|^2\mathrm{d}t$$

holds for $t \le s$. Combining this with (3.51), we obtain

$$\mathrm{d}\frac{|X(t)-Y(t)|^2}{\gamma_\theta(t)} \le \frac{2}{\gamma_\theta(t)}\langle(\sigma(t,X(t))-\sigma(t,Y(t)))(X(t)-Y(t)),\mathrm{d}\tilde{W}(t)\rangle \\ -\frac{|X(t)-Y(t)|^2}{\gamma_\theta(t)^2}\{2-K\,\gamma_\theta(t)+\gamma_\theta'(t)\}\mathrm{d}t \tag{3.55}$$

$$
\begin{aligned}
&= \frac{2}{\gamma_\theta(t)}\langle(\sigma(t,X(t))-\sigma(t,Y(t)))(X(t)-Y(t)),\mathrm{d}\tilde{W}(t)\rangle \\
&\quad -\frac{\theta}{\gamma_\theta(t)^2}|X(t)-Y(t)|^2\mathrm{d}t,\ t\le s.
\end{aligned}
$$

Multiplying by $\frac{1}{\theta}$ and integrating from 0 to s, we obtain

$$
\begin{aligned}
\int_0^s \frac{|X(t)-Y(t)|^2}{\gamma_\theta(t)^2}\mathrm{d}t &\le \int_0^s \frac{2}{\theta\gamma_\theta(t)}\langle(\sigma(t,X(t))-\sigma(t,Y(t)))(X(t)-Y(t)),\mathrm{d}\tilde{W}(t)\rangle \\
&\quad -\frac{|X(s)-Y(s)|^2}{\theta\gamma_\theta(s)}+\frac{|x-y|^2}{\theta\gamma_\theta(0)}.
\end{aligned}
$$

By Girsanov theorem, $\{\tilde{W}(t)\}_{t\le s}$ is a d-dimensional Brownian motion under the probability measure $\mathrm{d}\mathbb{Q}_s := R(s)\mathrm{d}\mathbb{P}$. So taking the expectation $\mathbb{E}_s$ with respect to $\mathbb{Q}_s$, we arrive at

$$
\mathbb{E}_s\int_0^s \frac{|X(t)-Y(t)|^2}{\gamma_\theta(t)^2}\mathrm{d}t \le \frac{|x-y|^2}{\theta\gamma_\theta(0)},\ s\in[0,T). \tag{3.56}
$$

By **(A3.5)** and the definitions of $R(t)$ and $\tilde{W}(t)$, we have

$$
\begin{aligned}
\log R(r) &= -\int_0^r \frac{1}{\gamma_\theta(t)}\langle\hat{\sigma}(t,X(t))(X(t)-Y(t)),\mathrm{d}\tilde{W}(t)\rangle \\
&\quad +\frac{1}{2}\int_0^r \frac{|\hat{\sigma}(t,X(t))(X(t)-Y(t))|^2}{\gamma_\theta(t)^2}\mathrm{d}t \\
&\le -\int_0^r \frac{1}{\gamma_\theta(t)}\langle\hat{\sigma}(t,X(t))(X(t)-Y(t)),\mathrm{d}\tilde{W}(t)\rangle \\
&\quad +\frac{1}{2\lambda}\int_0^r \frac{|X(t)-Y(t)|^2}{\gamma_\theta(t)^2}\mathrm{d}t,\ r\le s.
\end{aligned}
$$

Since $\{\tilde{W}(t)\}_{t\in[0,s]}$ is a d-dimensional Brownian motion under $\mathbb{Q}_s$, combining this with (3.56), we obtain

$$
\mathbb{E}\{R(s)\log R(s)\} = \mathbb{E}_s\log R(s) \le \frac{|x-y|^2}{2\lambda\theta\gamma_\theta(0)},\ s\in[0,T).
$$

By the martingale convergence theorem, we conclude that $\{R(s):\ s\in[0,T]\}$ is a well-defined martingale with

$$
\mathbb{E}\{R(s)\log R(s)\} \le \frac{|x-y|^2}{2\lambda\theta\gamma_\theta(0)} = \frac{K|x-y|^2}{2\lambda\theta(2-\theta)(1-\mathrm{e}^{-KT})},\ s\in[0,T].
$$

□

Lemma 3.4.2 ensures that under $\mathrm{d}\mathbb{Q} := R(T)\mathrm{d}\mathbb{P}$, $\{\tilde{W}(t)\}_{t\in[0,T]}$ is a Brownian motion, and

$$\mathbb{E}_{\mathbb{Q}}\{R(T)\log R(T)\} \le \frac{K|x-y|^2}{2\lambda\theta(2-\theta)(1-e^{-KT})}. \tag{3.57}$$

Then by (3.49) and (3.54), the coupling $(X(t),Y(t))$ is well constructed under $\mathbb{Q}$ for $t\in[0,T]$. Since $\int_0^T \gamma_\theta(t)^{-1}dt=\infty$, we shall see that the coupling is successful up to time T, so that $X(T)=Y(T)$ holds $\mathbb{Q}$-a.s. According to Theorem 1.1.1, (3.57) implies the log-Harnack inequality, and we will derive the Harnack inequality with power, provided that $\mathbb{E}R(T)^{\frac{p}{p-1}}<\infty$.

Lemma 3.4.3 *Assume* **(A3.4)–(A3.5)** *and* (3.48). *Then*

$$\sup_{s\in[0,T]}\mathbb{E}\left\{R(s)\exp\left[\frac{\theta^2}{8\delta^2}\int_0^s\frac{|X(t)-Y(t)|^2}{\gamma_\theta(t)^2}dt\right]\right\} \le \exp\left[\frac{\theta K\,|x-y|^2}{4\delta^2(2-\theta)(1-e^{-K\,T})}\right]. \tag{3.58}$$

Consequently,

$$\sup_{s\in[0,T]}\mathbb{E}R(s)^{1+\alpha}\le\exp\left[\frac{\theta K\,(2\delta+\theta\sqrt{\lambda})|x-y|^2}{8\delta^2(2-\theta)(\delta+\theta\sqrt{\lambda})(1-e^{-K\,T})}\right] \tag{3.59}$$

holds for

$$\alpha=\frac{\lambda\theta^2}{4\delta^2+4\theta\sqrt{\lambda}\delta}.$$

Proof. Let $\tau_n=\inf\{t\ge0:\int_0^t\frac{|X(s)-Y(s)|^2}{\gamma_\theta(s)^2}ds\ge n\}$. By (3.56), we have $\lim_{n\to\infty}\tau_n\ge T$, $\mathbb{Q}$-a.s. By (3.55), for any $r>0$, we have

$$\begin{aligned}\mathbb{E}_s\exp\left[r\int_0^{s\wedge\tau_n}\frac{|X(t)-Y(t)|^2}{\gamma_\theta(t)^2}dt\right] &\le \mathbb{E}_s\exp\left[\frac{r|x-y|^2}{\theta\gamma_\theta(0)}\right.\\ &\quad\left.+\frac{2r}{\theta}\int_0^{s\wedge\tau_n}\frac{1}{\gamma_\theta(t)}\langle(\sigma(t,X(t))-\sigma(t,Y(t)))(X(t)-Y(t)),d\tilde{W}(t)\rangle\right]\\ &\le\exp\left[\frac{rK\,|x-y|^2}{\theta(2-\theta)(1-e^{-K\,T})}\right]\left(\mathbb{E}_s\exp\left[\frac{8r^2\delta^2}{\theta^2}\int_0^{s\wedge\tau_n}\frac{|X(t)-Y(t)|^2}{\gamma_\theta(t)^2}dt\right]\right)^{\frac12},\end{aligned}$$

where the last step is due to (3.48) and the fact that $\mathbb{E}_s e^{M(t)}\le(\mathbb{E}_s e^{2\langle M\rangle(t)})^{\frac12}$ for a continuous exponential integrable $\mathbb{Q}_s$-martingale $M(t)$. Taking $r=\frac{\theta^2}{8\delta^2}$, we arrive at

$$\mathbb{E}_s\exp\left[\frac{\theta^2}{8\delta^2}\int_0^{s\wedge\tau_n}\frac{|X(t)-Y(t)|^2}{\gamma_\theta(t)^2}dt\right]\le\exp\left[\frac{\theta K\,|x-y|^2}{4\delta^2(2-\theta)(1-e^{-K\,T})}\right],\quad n\ge1.$$

This implies (3.58) by letting $n\to\infty$.

Next, by **(A3.4)** and the definition of $R(s)$, we have

$$\mathbb{E}R(s)^{1+\alpha}=\mathbb{E}_sR(s)^{\alpha}$$

$$= \mathbb{E}_s \exp\left[-\alpha \int_0^s \frac{1}{\gamma_\theta(t)} \langle \hat{\sigma}(t,X(t))(X(t)-Y(t)), \mathrm{d}\tilde{W}(t)\rangle \right. \tag{3.60}$$
$$\left. + \frac{\alpha}{2}\int_0^s \frac{|\hat{\sigma}(t,X(t))(X(t)-Y(t))|^2}{\gamma_\theta(t)^2}\mathrm{d}t\right].$$

Noting that for any exponential integrable martingale $M(t)$ with respect to $\mathbb{Q}_s$, one has

$$\begin{aligned}
&\mathbb{E}_s \exp\left[\alpha M(t) + \frac{\alpha}{2}\langle M\rangle(t)\right] \\
&= \mathbb{E}_s \exp\left[\alpha M(t) - \frac{\alpha^2 q}{2}\langle M\rangle(t) + \frac{\alpha(q\alpha+1)}{2}\langle M\rangle(t)\right] \\
&\le \left(\mathbb{E}_s \exp\left[\alpha q M(t) - \frac{\alpha^2 q^2}{2}\langle M\rangle(t)\right]\right)^{\frac{1}{q}} \\
&\quad \times \left(\mathbb{E}_s \exp\left[\frac{\alpha q(\alpha q+1)}{2(q-1)}\langle M\rangle(t)\right]\right)^{\frac{q-1}{q}} \\
&= \left(\mathbb{E}_s \exp\left[\frac{\alpha q(\alpha q+1)}{2(q-1)}\langle M\rangle(t)\right]\right)^{\frac{q-1}{q}}, \quad q>1,
\end{aligned}$$

it follows from (3.60) and **(A3.5)** that

$$\mathbb{E}R(s)^{1+\alpha} \le \left(\mathbb{E}_s \exp\left[\frac{q\alpha(q\alpha+1)}{2(q-1)\lambda}\int_0^s \frac{|X(t)-Y(t)|^2}{\gamma_\theta(t)^2}\mathrm{d}t\right]\right)^{\frac{q-1}{q}}. \tag{3.61}$$

Take

$$q = 1+\sqrt{1+\alpha^{-1}}, \tag{3.62}$$

which minimizes $\frac{q(q\alpha+1)}{q-1}$ such that

$$\begin{aligned}
\frac{q\alpha(q\alpha+1)}{2\lambda(q-1)} &= \frac{\alpha+\sqrt{\alpha(\alpha+1)}}{2\lambda\sqrt{1+\alpha^{-1}}}(\alpha+1+\sqrt{\alpha(\alpha+1)}) \\
&= \frac{(\alpha+\sqrt{\alpha^2+\alpha})^2}{2\lambda} = \frac{\theta^2}{8\delta^2}.
\end{aligned} \tag{3.63}$$

Combining (3.61) with (3.58) and (3.63), and noting that due to (3.62) and the definition of α,

$$\frac{q-1}{q} = \frac{\sqrt{1+\alpha^{-1}}}{1+\sqrt{1+\alpha^{-1}}} = \frac{2\delta+\theta\sqrt{\lambda}}{2\delta+2\theta\sqrt{\lambda}},$$

we obtain

$$\mathbb{E}R(s)^{1+\alpha} \le \exp\left[\frac{\theta K\,(2\delta+\theta\sqrt{\lambda})|x-y|^2}{8\delta^2(2-\theta)(\delta+\theta\sqrt{\lambda})(1-\mathrm{e}^{-K\,T})}\right].$$

This completes the proof. □

3.4.2 Proof of Theorem 3.4.1

As explained at the beginning of this section, we consider only the case that $\mathbb{H} = \mathbb{R}^d$.

(1) By Lemma 3.4.2, $\{R(s)\}_{s\in[0,T]}$ is a uniformly integrable martingale and $\{\tilde{W}(t)\}_{t\leq T}$ is a d-dimensional Brownian motion under the probability $\mathbb{Q}$. Thus, $Y(t)$ can be solved from (3.54) up to time T. Let

$$\tau = \inf\{t \in [0,T] : X(t) = Y(t)\}$$

and set $\inf\emptyset = \infty$ by convention. We claim that $\tau \leq T$ and thus $X(T) = Y(T)$, $\mathbb{Q}$-a.s. Indeed, if for some $\omega \in \Omega$ we have $\tau(\omega) > T$, then by the continuity of the processes, we have

$$\inf_{t\in[0,T]} |X(t) - Y(t)|^2(\omega) > 0.$$

So

$$\int_0^T \frac{|X(t)-Y(t)|^2}{\gamma_\theta(t)^2} \mathrm{d}t = \infty$$

holds on the set $\{\tau > T\}$. But according to (3.56), we have

$$\mathbb{E}_{\mathbb{Q}} \int_0^T \frac{|X(t)-Y(t)|^2}{\gamma_\theta(t)^2} \mathrm{d}t < \infty,$$

and we conclude that $\mathbb{Q}(\tau > T) = 0$, i.e., $\tau \leq T$ and $X(T) = Y(T)$ $\mathbb{Q}$-a.s. Therefore, according to Theorem 1.1.1, Theorem 3.4.1(1) follows from (3.57) with $\theta = 1$.

(2) Take $\theta = \frac{2\delta}{\sqrt{\lambda}(\sqrt{p}-1)}$, which is in $(0,2)$ for $p > (1+\frac{\delta}{\sqrt{\lambda}})^2$. We have

$$\frac{p}{p-1} = 1 + \frac{\lambda\theta^2}{4\delta(\delta+\theta\sqrt{\lambda})} = 1+\alpha.$$

So Lemma 3.4.3 yields that

$$\begin{aligned}
(\mathbb{E}R(T)^{\frac{p}{p-1}})^{p-1} &= (\mathbb{E}R(T)^{1+\alpha})^{p-1} \\
&\leq \exp\Big[\frac{(p-1)\theta K\,(2\delta+\theta\sqrt{\lambda})|x-y|^2}{8\delta^2(2-\theta)(\delta+\theta\sqrt{\lambda})(1-\mathrm{e}^{-K\,T})}\Big] \\
&= \exp\Big[\frac{K\,\sqrt{p}(\sqrt{p}-1)|x-y|^2}{4\delta[(\sqrt{p}-1)\sqrt{\lambda}-\delta](1-\mathrm{e}^{-K\,T})}\Big].
\end{aligned}$$

According to Theorem 1.1.1, this implies the desired Harnack inequality in (2) if $\delta_p = \delta$. When $\delta_p > \delta$, (3.48) holds for δ_p in place of δ, so that the desired Harnack inequality remains true.

3.5 Multiplicative Noise: Bismut Formula

In this section, we establish the Bismut derivative formula for equations with multiplicative noise using the Malliavin calculus; see Sect. 4.3.3 for the study of stochastic functional differential equations. To this end, we assume the following:

(A3.6) There exists a constant $K \geq 0$ such that for every $v \in \mathbb{H}$,

$$2 \sup_{t\in[0,T],x\in\mathbb{H}} \Big\{|v|\cdot|\nabla_v b(t,\cdot)(x)| + \|\nabla_v \sigma(t,\cdot)(x)\|_{HS}^2\Big\} \leq K|v|^2.$$

Theorem 3.5.1 *Assume* **(A3.1)**, **(A3.2)**, **(A3.5)**, *and* **(A3.6)**. *Let* $v \in \mathbb{H}$, $\theta \in (0,2)$, *and* γ_θ *be in* (3.50).

(1) *The equation*

$$\begin{aligned} \mathrm{d}v(t) = &\Big\{Av(t) + \nabla_{v(t)} b(t,\cdot)(X(t)) - \frac{v(t)}{\gamma_\theta(t)}\Big\}\mathrm{d}t \\ &+ \nabla_{v(t)}\sigma(t,\cdot)(X(t))\mathrm{d}W(t), \ \ v(0) = v, t \in [0,T) \end{aligned} \tag{3.64}$$

has a unique solution in the sense of Definition 2.1.1, and the solution satisfies $v(T) := \lim_{t\to T} v(t) = 0$ *and*

$$\mathbb{E}\int_0^T \frac{|v(t)|^2}{\gamma_\theta(t)^2}\mathrm{d}t \leq \frac{|v|^2}{\theta\gamma_\theta(0)}, \tag{3.65}$$

$$\mathbb{E}\Big(\frac{|v(t)|^2}{\gamma_\theta(t)}\Big) \leq \frac{|v|^2}{\gamma_\theta(0)}\mathrm{e}^{-\int_0^t \frac{\theta}{\gamma_\theta(s)}\mathrm{d}s}, \ \ t \in [0,T]. \tag{3.66}$$

(2) *For every* $f \in \mathscr{B}_b(\mathbb{H})$,

$$\nabla_v P_T f = \mathbb{E}\bigg(f(X(T))\int_0^T \frac{1}{\gamma_\theta(t)}\langle\hat{\sigma}(t,X(t))^{-1}v(t), \mathrm{d}W(t)\rangle\bigg).$$

Consequently,

$$|\nabla P_T f|^2 \leq \frac{K}{\lambda(1-\mathrm{e}^{-KT})}P_T f^2.$$

Proof.(1) It is easy to see that our assumptions imply **(A2.1)**–**(A2.4)** for $\mathbb{V} = \mathbb{H}$ and $t \in [0,T)$. Therefore, (3.64) has a unique solution $\{v(t)\}_{t\in[0,T)}$. By **(A3.6)** and Itô's formula for $|v(t)|^2$, we have

$$\mathrm{d}|v(t)|^2 \leq \Big(K - \frac{2}{\gamma_\theta(t)}\Big)|v(t)|^2\mathrm{d}t + 2\langle v(t), \nabla_{v(t)}\sigma(t,\cdot)(X(t))\mathrm{d}W(t)\rangle, \ \ t < T.$$

Then, due to (3.51),

$$\mathrm{d}\frac{|v(t)|^2}{\gamma_\theta(t)} \leq \ (K\gamma_\theta(t) - \gamma_\theta'(t) - 2)\frac{|v(t)|^2}{\gamma_\theta(t)^2}\mathrm{d}t + \frac{2}{\gamma_\theta(t)}\langle v(t), \nabla_{v(t)}\sigma(t,\cdot)(X(t))\mathrm{d}W(t)\rangle$$

$$= -\frac{\theta|v(t)|^2}{\gamma_\theta(t)^2}\mathrm{d}t + \frac{2}{\gamma_\theta(t)}\langle v(t), \nabla_{v(t)}\sigma(t,\cdot)(X(t))\mathrm{d}W(t)\rangle, \quad t < T.$$

This implies (3.65) and (3.66). Moreover, by the Burkholder–Davis–Gundy inequality, **(A3.6)**, and (3.65), this also implies

$$\mathbb{E}\sup_{t\in[0,T)}\frac{|v(t)|^2}{\gamma_\theta(t)} \le C\left(\mathbb{E}\int_0^T \frac{|v(t)|^2}{\gamma_\theta(t)^2}\mathrm{d}t\right)^{\frac{1}{2}} < \infty$$

for some constant $C > 0$. Since $\gamma_\theta(t) \downarrow 0$ as $t \uparrow T$, we conclude that $v(T) := \lim_{t\to T} v(t) = 0$.

(2) Now let

$$h(t) = \int_0^t \frac{1}{\gamma_\theta(s)}\hat{\sigma}(s,X(s))^{-1}v(s)\mathrm{d}s, \quad s \in [0,T].$$

Then h is adapted, and by **(A3.6)** and (3.65),

$$\mathbb{E}\int_0^T |h'(t)|^2\mathrm{d}t < \infty. \tag{3.67}$$

It is easy to see that $Z(t) := \nabla_v X(t) - D_h X(t)$ solves the equation

$$\mathrm{d}Z(t) = \Big\{AZ(t) + \nabla_{Z(t)}b(t,\cdot)(X(t)) - \frac{v(t)}{\gamma_\theta(t)}\Big\}\mathrm{d}t + \nabla_{Z(t)}\sigma(t,\cdot)(X(t))\mathrm{d}W(t)$$

with $Z(0) = v$. Therefore, $\Gamma(t) := Z(t) - v(t)$ solves the equation

$$\mathrm{d}\Gamma(t) = \{A\Gamma(t) + \nabla_{\Gamma(t)}b(t,\cdot)(X(t))\}\mathrm{d}t + \nabla_{\Gamma(t)}\sigma(t,\cdot)(X(t))\mathrm{d}W(t), \quad \Gamma(0) = 0.$$

By the uniqueness of the solution to this equation, we have $\Gamma(t) = 0$, i.e., $Z(t) = v(t)$. In particular, $Z(T) := \lim_{t\to T} Z(t) = 0$; that is, $\nabla_v X(T) = D_h X(T)$. Therefore, the desired Bismut derivative formula follows from Theorem 1.2.1 by noting that the adapted property and (3.67) imply $h \in \mathscr{D}(D^*)$ and

$$\begin{aligned} D^*h &= \int_0^T \langle h'(t), \mathrm{d}W(t)\rangle \\ &= \int_0^T \frac{1}{\gamma_\theta(t)}\langle \hat{\sigma}(t,X(t))^{-1}v(t), \mathrm{d}W(t)\rangle. \end{aligned} \tag{3.68}$$

Finally, to derive the gradient estimate, we take $\theta = 1$. By the Bismut formula and using (3.65), we obtain

$$|\nabla_v P_T f|^2 \le \frac{P_T f^2}{\lambda}\mathbb{E}\int_0^T \frac{|v(t)|^2}{\gamma_1(t)^2}\mathrm{d}t \le \frac{|v|^2 K P_T f^2}{\lambda(1-\mathrm{e}^{-KT})}.$$

□

Chapter 4
Stochastic Functional (Partial) Differential Equations

4.1 Solutions and Finite-Dimensional Approximations

Let $r_0 > 0$ be a fixed constant. For a metric space E, let $\mathscr{C}(E) = C([-r_0, 0]; E)$. When $E = \mathbb{R}^d$, we simply set $\mathscr{C} = \mathscr{C}(E)$. For a continuous process $(Z(t))_{t \geq -r_0}$ on E, we will denote $(Z_t)_{t \geq 0}$, the corresponding functional process on $\mathscr{C}(E)$, by letting

$$Z_t(s) = Z(t+s), \quad s \in [-r_0, 0].$$

4.1.1 Stochastic Functional Differential Equations

In this subsection we consider SDDEs on $\mathbb{R}^d$. The existence and uniqueness result on the solution we are going to introduce will be also used for construction of couplings. For this purpose, one has only to solve a joint equation up to the coupling time, i.e., the exit time of the solution from the off-diagonal $D := \{(x,y) : x \neq y\}$. In general, we consider solutions up to the exit time of an open set $D \subset \mathbb{R}^d$. For fixed $\mathbf{T} \in (0, \infty]$ and a nonempty open domain D in $\mathbb{R}^d$, we consider the SDDE

$$\mathrm{d}X(t) = \bar{b}(t, X_t)\mathrm{d}t + \bar{\sigma}(t, X_t)\mathrm{d}B(t), \quad X_0 \in \mathscr{C}(D), \tag{4.1}$$

where $B(t)$ is an m-dimensional Brownian motion, $\bar{b} : [0, \mathbf{T}) \times \mathscr{C}(D) \to \mathbb{R}^d$ and σ : $[0, \mathbf{T}) \times \mathscr{C}(D) \to \mathbb{R}^d \otimes \mathbb{R}^m$ are measurable, bounded on $[0, t] \times \mathscr{C}(\mathbf{K})$ for $t \in [0, \mathbf{T})$ and a compact set $\mathbf{K} \subset D$, and continuous in the second variable.

A continuous adapted process $\{X(t)\}_{t \in [0, \zeta)}$ with $X_0 \in \mathscr{C}(D)$ and lifetime

$$\zeta := \lim_{n \to \infty} \mathbf{T} \wedge \inf\{t \in [0, \mathbf{T}) : \mathrm{dist}(X(t), \partial D) \leq n^{-1}, |X(t)| \geq n\}$$

is called a strong solution to the equation if a.s.,

F.-Y. Wang, *Harnack Inequalities for Stochastic Partial Differential Equations*, SpringerBriefs in Mathematics, DOI 10.1007/978-1-4614-7934-5_4,

$$X(t) = X(0) + \int_0^t b(s,X_s)\mathrm{d}s + \int_0^t \bar{\sigma}(s,X_s)\mathrm{d}B(s), \quad t \in [0,\zeta).$$

In this case, $(X_t)_{t\geq 0}$ is called the segment (or functional) solution to the equation.

To characterize the non-Lipschitz regularity of coefficients ensuring the existence of solutions as in [18, 25, 48] for the case without delay, we introduce the class of functions

$$\mathscr{U} := \left\{ u \in C^1((0,\infty),[1,\infty)) : \int_0^1 \frac{\mathrm{d}s}{su(s)} = \infty, \ s \mapsto su(s) \text{ is increasing and concave} \right\}.$$

Theorem 4.1.1 *Assume that there exists a sequence of compact sets* $\mathbf{K}_n \uparrow D$ *such that for every* $n \geq 1$,

$$\begin{aligned} &2\langle \bar{b}(t,\xi) - \bar{b}(t,\eta), \xi(0) - \eta(0)\rangle + \|\sigma(t,\xi) - \sigma(t,\eta)\|_{HS}^2 \\ &\quad \leq \|\xi - \eta\|_\infty^2 u_n(\|\xi - \eta\|_\infty^2), \end{aligned} \tag{4.2}$$

$$\|\sigma(t,\xi) - \sigma(t,\eta)\|_{HS}^2 \leq \|\xi - \eta\|_\infty^2 u_n(\|\xi - \eta\|_\infty^2) \tag{4.3}$$

hold for some $u_n \in \mathscr{U}$ *and all* $\xi, \eta \in \mathscr{C}(\mathbf{K}_n)$, $t \leq \frac{n\mathbf{T}}{n+1} \wedge n$. *Then for any initial data* $X_0 \in \mathscr{C}(D)$, (4.1) *has a unique solution* $X(t)$ *up to its lifetime*

$$\zeta := \mathbf{T} \wedge \lim_{n\to\infty} \inf \{ t \in [0,\mathbf{T}) : X(t) \notin \mathbf{K}_n \}.$$

Proof. (a) We first assume that $D = \mathbb{R}^d$ and $\bar{b}, \sigma$ are bounded and continuous in the second variable, and prove the existence and uniqueness of the solution up to any time $T' < \mathbf{T}$. According to the Yamada–Watanabe principle [71], we shall verify below the existence of a weak solution and the pathwise uniqueness of the strong solution.

(a1) The existence of a weak solution. Let $\mathbf{B}(s) = \tilde{B}(r_0 + 1 + s)$, $s \in [-r_0, 0]$, where $\tilde{B}(s)$ is a d-dimensional Brownian motion with $\tilde{B}(0) = 0$. Define

$$\sigma_n(t,\xi) = \mathbb{E}\sigma(t,\xi + n^{-1}\mathbf{B}), \quad \bar{b}_n(t,\xi) = \mathbb{E}\bar{b}(t,\xi + n^{-1}\mathbf{B}), \quad n \geq 1.$$

Then for every $n \geq 1$, σ_n and $\bar{b}_n$ are Lipschitz continuous in the second variable uniformly in the first variable. According to the proof of [44, Theorem 2.3], the equation

$$\mathrm{d}X^{(n)}(t) = \bar{b}_n(t,X_t^{(n)})\mathrm{d}t + \bar{\sigma}_n(t,X_t^{(n)})\mathrm{d}B(t), \quad X_0^{(n)} = X_0,$$

has a unique strong solution $X^{(n)}$ up to time T'. To see that $X^{(n)}$ converges weakly as $n \to \infty$, we take the reference function

$$g_\varepsilon(h) := \|h\|_\infty + \sup_{s,t\in[0,T'], s\neq t} \frac{|h(s) - h(t)|}{|t-s|^\varepsilon}$$

for a fixed number $\varepsilon \in (0, \frac{1}{2})$. It is well known that g_ε is a compact function on $C([0,T'];\mathbb{R}^d)$, i.e., $\{g_\varepsilon \le r\}$ is compact under the uniform norm for every $r > 0$. Since $\bar{b}_n$ and σ_n are bounded uniformly in n, by a standard argument leading to the Hölder continuity of order $\varepsilon \in (0, \frac{1}{2})$ for the Brownian path, we have

$$\lim_{R\to\infty} \sup_{n\ge 1} \mathbb{P}(g_\varepsilon(X^{(n)}) \ge R) = 0.$$

Let $\mathbf{P}^{(n)}$ be the distribution of $X^{(n)}$. Then the family $\{\mathbf{P}^{(n)}\}_{n\ge1}$ is tight, and hence (up to a subsequence) converges weakly to a probability measure $\mathbf{P}$ on $\Omega := C([0,T'];\mathbb{R}^d)$. Let $\mathscr{F}_t = \sigma(\omega \mapsto \omega(s) : s \le t)$ for $t \in [0,T']$. Then the process

$$X(t)(\omega) := \omega(t),\ \ t \in [0,T'],\ \omega \in \Omega,$$

is $\mathscr{F}_t$-adapted. Since $\mathbf{P}^{(n)}$ is the distribution of $X^{(n)}$, we see that

$$M^{(n)}(t) := X(t) - \int_0^t \bar{b}_n(s,X_s)\mathrm{d}s,\ \ t \in [0,T'],$$

is a $\mathbf{P}^{(n)}$-martingale with

$$\langle M_i^{(n)}, M_j^{(n)}\rangle(t) = \sum_{i=1}^m \int_0^t \{(\sigma_n)_{ik}(\sigma_n)_{jk}\}(s,X_s)\mathrm{d}s,\ \ 1 \le i,j \le d.$$

Since $\sigma_n, \bar{b}_n$ are bounded uniformly in n, and as $n \to \infty$ converge to σ and $\bar{b}$ locally uniformly, by the weak convergence of $\mathbf{P}^{(n)}$ to $\mathbf{P}$ and letting $n \to \infty$, we see that

$$M(t) := X(t) - \int_0^t \bar{b}(s,X_s)\mathrm{d}s,\ \ s \in [0,T'],$$

is a $\mathbf{P}$-martingale with

$$\langle M_i, M_j\rangle(t) = \sum_{i=1}^m \int_0^t \{\bar{\sigma}_{ik}\sigma_{jk}\}(s,X_s)\mathrm{d}s,\ \ 1 \le i,j \le d.$$

According to [22, Theorem II.7.1], this implies

$$M(t) = \int_0^t \sigma(s,X_s)\mathrm{d}\tilde{B}(s),\ \ t \in [0,T'],$$

for some m-dimensional Brownian motion $\tilde{B}$ on the filtered probability space $(\Omega, \mathscr{F}_t, \mathbf{P})$. Therefore, the equation has a weak solution up to time T'.

(a2) The pathwise uniqueness. Let $X(t)$ and $Y(t)$ for $t \in [0,T']$ be two strong solutions with $X_0 = Y_0$. Let $Z = X - Y$ and

$$\tau_n = T' \wedge \inf\{t \in [0,T'] :\ |X(t)| + |Y(t)| \ge n\}.$$

By Itô's formula and (4.2), we have

$$\mathrm{d}|Z(t)|^2 \leq 2\langle(\sigma(t,X_t)-\sigma(t,Y_t))\mathrm{d}B(t), Z(t)\rangle + \|Z_t\|_\infty^2 u_n(\|Z_t\|_\infty^2), \ t \leq \tau_n. \quad (4.4)$$

Let

$$\ell_n(t) := \sup_{s \leq t\wedge\tau_n} |Z(s)|^2, \ t \geq 0.$$

Noting that $su_n(s)$ is increasing in s, we have

$$\|Z_t\|_\infty^2 u_n(\|Z_t\|_\infty^2) \leq \ell_n(t) u_n(\ell_n(t)), \ t \geq 0.$$

So by (4.3), (4.4), and the Burkholder–Davis–Gundy inequality, there exist constants $C_1, C_2 > 0$ such that

$$\begin{aligned}\mathbb{E}\ell_n(t) &\leq \int_0^t \mathbb{E}\ell_n(s)u_n(\ell_n(s))\mathrm{d}s + C_1\mathbb{E}\left(\ell_n(t)\int_0^t \ell_n(s)u_n(\ell_n(s))\mathrm{d}s\right)^{\frac{1}{2}} \\ &\leq \frac{1}{2}\mathbb{E}\ell_n(t) + C_2\int_0^t \mathbb{E}\ell_n(s)u_n(\ell_n(s))\mathrm{d}s.\end{aligned}$$

Since $s \mapsto su_n(s)$ is concave, due to Jensen's inequality this implies that

$$\mathbb{E}\ell_n(t) \leq 2C_2\int_0^t \mathbb{E}\ell_n(s)u_n\big(\mathbb{E}\ell_n(s)\big)\mathrm{d}s.$$

Let $G(s) = \int_1^s \frac{1}{ru_n(r)}\mathrm{d}r$, $s > 0$, and let G^{-1} be the inverse of G with domain $G([0,\infty))$. Since $\int_0^1 \frac{1}{su_n(s)}\mathrm{d}s = \infty$, we have $[-\infty, 0] \subset G([0,\infty))$ with $G^{-1}(-\infty) = 0$. Then by Bihari's inequality (cf. [33, Theorem 1.8.2]), we obtain

$$\mathbb{E}\ell_n(t) \leq G^{-1}\big(G(0) + G(2C_2t)\big) = G^{-1}(-\infty) = 0.$$

This implies that $X(t) = Y(t)$ for $t \leq \tau_n$ for $n \geq 1$. Since $\bar{b}$ and σ are bounded, we have $\tau_n \uparrow T'$. Therefore, $X(t) = Y(t)$ for $t \in [0, T']$.

(b) In general, for each $n \geq 1$, we may find a Lipschitz continuous function h_n on $\mathscr{C}$ and a compact set $\mathbf{K} \subset D$ such that $h_n|_{\mathscr{C}(\mathbf{K}_n)} = 1$ and $h_n|_{(\mathscr{C}(\mathbf{K}))^c} = 0$. Let

$$\bar{b}_n(t,\cdot) = h_n(\cdot)b(t,\cdot), \ \ \sigma_n(t,\cdot) = h_n(\cdot)\sigma(t,\cdot).$$

Let $T_n = \frac{n\mathbf{T}}{n+1} \wedge n$. Then for all $n \geq 1$, $\bar{b}_n$ and $\bar{\sigma}_n$ are bounded on $[0,T_n] \times \mathscr{C}(D)$ and continuous in the second variable. According to (a), the equation

$$\mathrm{d}X^{(n)}(t) = \bar{b}_n(t, X_t^{(n)})\mathrm{d}t + \bar{\sigma}_n(t, X_t^{(n)})\mathrm{d}B(t), \ X_0^{(n)} = X_0, \quad (4.5)$$

has a unique strong solution $\{X^{(n)}(t)\}_{t\in[0,\mathbf{T})}$. Since $h_n = 1$ on $\mathscr{C}(\mathbf{K}_n)$, so that (4.5) coincides with (4.1) before the solution leaves $\mathbf{K}_n$, (4.1) has a unique solution $X(t)$ up to the time

$$\zeta_n := T_n \wedge \inf\{t \in [0,T_n] : X(t) \notin \mathbf{K}_n\}.$$

Therefore, (4.1) has a unique solution up to the lifetime ζ. □

To ensure the existence and nonexplosion of global solutions, we make the following assumption.

(A4.1) $D = \mathbb{R}^d$, $\mathbf{T} = \infty$, and there exist $u \in \mathscr{U}$ and a positive function $H \in C([0,\infty))$ such that for every $\xi, \eta \in \mathscr{C}$, and $t \geq 0$,

$$\begin{aligned}
&2\langle \bar{b}(t,\xi) - \bar{b}(t,\eta), \xi(0) - \eta(0)\rangle + \|\sigma(t,\xi) - \sigma(t,\eta)\|_{HS}^2 \\
&\quad \leq H(t)\|\xi - \eta\|_\infty^2 u(\|\xi - \eta\|_\infty^2), \\
&\|\sigma(t,\xi) - \sigma(t,\eta)\|_{HS}^2 \leq H(t)\|\xi - \eta\|_\infty^2 u(\|\xi - \eta\|_\infty^2), \\
&2\langle \bar{b}(t,\xi), \xi(0)\rangle\rangle + \|\sigma(t,\xi)\|_{HS}^2 \leq H(t) + H(t)\|\xi\|_\infty^2 u(\|\xi\|_\infty^2).
\end{aligned}$$

Since by Itô's formula it is easy to see that **(A4.1)** implies (4.2) and (4.3) for $u_n = H(n)u$ and the nonexplosion of the solution, the following is a direct consequence of Theorem 4.1.1.

Corollary 4.1.2 *Assume* **(A4.1)**. *Then for every $\mathscr{F}_0$-measurable X_0, (4.1) has a unique strong solution, and the solution is nonexplosive, i.e., with lifetime $\zeta = \infty$.*

Now, assuming **(A4.1)**, for every $\xi \in \mathscr{C}$, let X_t^ξ be the unique segment solution of (4.1) with $X_0 = \xi$. We shall investigate Harnack/shift Harnack inequalities and derivative formulas for the associated semigroup P_T:

$$P_T f(\xi) := \mathbb{E} f(X_T^\xi), \quad f \in \mathscr{B}_b(\mathscr{C}), \xi \in \mathscr{C}.$$

4.1.2 Semilinear Stochastic Functional Partial Differential Equations

As in the case without delay considered in Chap. 3, we may extend results derived in finite dimensions to semilinear stochastic functional partial differential equations (SDPDEs). Let $\mathbb{H}, \tilde{\mathbb{H}}$, $(A, \mathscr{D}(A))$, $S(t)$, $W(t)$, and σ be as introduced at the beginning of Sect. 3.1. Let $\mathscr{C}(\mathbb{H}) = C([-r_0,0];\mathbb{H})$, and let

$$b : [0,T] \times \mathscr{C}(\mathbb{H}) \to \tilde{\mathbb{H}}, \quad a : [0,T] \times \mathbb{H} \to \mathbb{H},$$

be measurable. Assume the following:

(A4.2) For every $s > 0$ and $t \in [0,T]$, $S(s)b(t,0) \in \mathbb{H}$, and there exists $\varepsilon > 0$ such that

$$\int_0^T \sup_{r \in [0,T]} |S(s)b(r,0)|^{2(1+\varepsilon)} ds < \infty,$$

and there exists a positive function $K_b \in C((0,T])$ with $\int_0^t K_b(s)^{1+\varepsilon} ds < \infty$ such that

$$|S(t)(b(s,\xi) - b(s,\eta))|^2 \leq K_b(t)\|\xi - \eta\|_\infty^2, \quad s,t \in [0,T], \xi, \eta \in \mathbb{H}.$$

(A4.3) There exists $\varepsilon > 0$ such that $\int_0^T \sup_{r\in[0,T]} \|S(s)\sigma(r,0)\|_{HS}^{2(1+\varepsilon)} ds < \infty$, and there exists a positive function $K_\sigma \in C((0,T])$ such that $\int_0^T K_\sigma(s)^{1+\varepsilon} ds < \infty$ and

$$\|S(t)(\sigma(s,x)-\sigma(s,y))\|_{HS}^2 \le K_\sigma(t)|x-y|^2, \quad s,t \in [0,T], x,y \in \mathbb{H}.$$

Consider the semilinear SDPDE

$$\mathrm{d}X(t) = \big\{AX(t) + a(t,X(t)) + b(t,X_t)\big\}\mathrm{d}t + \sigma(t,X(t))\mathrm{d}W(t), \quad t \in [0,T]. \tag{4.6}$$

As in Definition 3.1.1, a progressively measurable process on $\mathbb{H}$ is called a mild solution to (4.6) if for every $t \in [0,T]$,

$$\int_0^t \mathbb{E}\big(|S(t-s)\{b(s,X_s)+a(s,X(s))\}| + \|S(t-s)\sigma(s,X(s))\|_{HS}^2\big)\mathrm{d}s < \infty$$

and almost surely

$$X(t) = S(t)X(0) + \int_0^t S(t-s)\{b(s,X_s)+a(s,X(s))\}\mathrm{d}s + \int_0^t S(t-s)\sigma(s,X(s))\mathrm{d}W(s).$$

When **(A3.3)** holds, let $\mathbb{H}_n$, A_n, b_n, σ_n, W_n, and $\mathscr{P}_n$ be as in Sect. 3.1, and let $Z_n = \mathscr{P}_n Z$. Consider the following finite-dimensional equations on $\mathbb{H}_n$ for $n \ge 1$:

$$\mathrm{d}X^{(n)}(t) = \big\{A_n X^{(n)}(t) + Z_n(t,X^{(n)}(t)) + b_n(t,X_t^{(n)})\big\}\mathrm{d}t + \sigma_n(t,X^{(n)}(t))\mathrm{d}W(t).$$

The following result can be easily proved by modifying the proofs of Theorems 3.1.1 and 3.1.2 using **(A4.2)** and **(A4.3)** instead of **(A3.1)** and **(A3.2)**.

Theorem 4.1.3 *Assume* **(A4.2)** *and* **(A4.3)** *for some* $\varepsilon > 0$. *If there exists* $\alpha \in (0,1)$ *such that* (3.6) *holds, then there exists a constant* $C > 0$ *such that for every* $X_0 \in L^{2(1+\varepsilon)}(\Omega \to \mathscr{C}(\mathbb{H}), \mathscr{F}_0, \mathbb{P})$, (4.6) *has a unique mild solution that is continuous and satisfies*

$$\mathbb{E}\sup_{[0,T]} \|X_s\|_\infty^{2(1+\varepsilon)} \le C(1 + \mathbb{E}\|X_0\|_\infty^{2(1+\varepsilon)}).$$

If, moreover, **(A3.3)** *holds, then*

$$\lim_{n\to\infty} \mathbb{E}\|X_t^{(n)} - X_t\|_\infty^2 = 0, \quad t \in [0,T].$$

As in the finite-dimensional case, we shall investigate

$$P_T f(\xi) := \mathbb{E} f(X_T^\xi), \quad f \in \mathscr{B}_b(\mathscr{C}(\mathbb{H})), \; \xi \in \mathscr{C}(\mathbb{H}).$$

4.2 Elliptic Stochastic Functional Partial Differential Equations with Additive Noise

Using coupling by change of measures, the Harnack inequality with powers for SDDEs with additive noise was first investigated in [15]. In this section we also study the derivative formula and shift Harnack inequalities. Consider the following equation on $\mathbb{R}^d$:

$$dX(t) = \{a(t,X(t)) + b(t,X_t)\}dt + \sigma(t)dB(t), \quad X_0 \in \mathscr{C}, t \in [0,T]. \tag{4.7}$$

Assume that $\bar{b}(t,\xi) := a(t,\xi(0)) + b(t,\xi)$ and $\sigma(t,\xi) = \sigma(t)$ satisfy **(A4.1)**. Let P_t be the Markov semigroup for the segment solution.

4.2.1 Harnack Inequalities and Bismut Formula

Theorem 4.2.1 *Assume* **(A4.1)** *for* $\bar{b}(t,\xi) := a(t,\xi(0)) + b(t,\xi)$. *Let* $K \in \mathbb{R}$ *and* $K', \lambda > 0$ *be such that*

$$\langle a(t,x) - a(t,y), x-y\rangle \le K|x-y|^2, \quad \|\sigma(s)^{-1}x\|^2 \le \frac{\|x\|^2}{\lambda}, \quad x,y \in \mathbb{R}^d, \tag{4.8}$$

$$|b(t,\xi) - b(t,\eta)| \le K'\|\xi - \eta\|_\infty, \quad \xi,\eta \in \mathscr{C}. \tag{4.9}$$

Then for every positive $f \in \mathscr{B}_b(\mathscr{C})$ *and* $T > r_0$,

$$P_T \log f(\eta) \le \log P_T f(\xi) + C(T,\xi,\eta)^2, \quad \xi,\eta \in \mathscr{C};$$

$$(P_T f(\eta))^p \le (P_T f^p(\xi)) \exp\left[\frac{pC(T,\xi,\eta)^2}{2(p-1)}\right], \quad p > 1, \xi,\eta \in \mathscr{C},$$

where

$$\begin{aligned} C(T,\xi,\eta) := {} & K'\sqrt{r_0}\|\xi-\eta\|_\infty + |\xi(0)-\eta(0)|\left\{\left(\frac{2K}{\lambda(1-e^{-2K(T-r_0)})}\right)^{\frac{1}{2}}\right. \\ & \left. + K'\left(\frac{1-e^{-4K(T-r_0)} - 4K(T-r_0)e^{-2K(T-r_0)}}{2K(1-e^{-2K(T-r_0)})^2}\right)^{\frac{1}{2}}\right\}. \end{aligned}$$

Proof. Let $X(t)$ solve (4.7) for $X_0 = \xi$, and let $Y(t)$ solve the equation

$$\begin{aligned} dY(t) = {} & \left\{a(t,Y(t)) + b(t,X_t) + \gamma(t)1_{[0,\tau)}(t)\frac{X(t)-Y(t)}{|X(t)-Y(t)|}\right\}dt \\ & + \sigma(t)dB(t), \quad Y_0 = \eta, \end{aligned} \tag{4.10}$$

where $\tau := T \wedge \inf\{t \ge 0 : X(t) = Y(t)\}$ is the coupling time and

$$\gamma(t) = \frac{2K\mathrm{e}^{-Kt}|\xi(0)-\eta(0)|}{1-\mathrm{e}^{-2K(T-r_0)}}, \quad t \geq 0. \tag{4.11}$$

Due to (4.8) and (4.9), we are able to apply Theorem 4.1.1 with $D = \{(x,y) : x \neq y\} \subset \mathbb{R}^d \times \mathbb{R}^d$ and $u(r) = (K \vee K')$, so that the coupling $(X(t),Y(t))$ is well defined for $t < \tau$. Moreover, (4.8) implies that

$$\mathrm{d}|X(t)-Y(t)| \leq \{K|X(t)-Y(t)| - \gamma(t)\}\mathrm{d}t, \quad t \in [0,\tau).$$

Then

$$\begin{aligned} |X(t)-Y(t)| &\leq \mathrm{e}^{Kt}\left(|\xi(0)-\eta(0)| - \int_0^t \mathrm{e}^{-Ks}\gamma(s)\mathrm{d}s\right) \\ &= \mathrm{e}^{Kt}|\xi(0)-\eta(0)|\frac{\mathrm{e}^{-2Kt}-\mathrm{e}^{-2K(T-r_0)}}{1-\mathrm{e}^{-2K(T-t_0)}}, \quad t \in [0,\tau). \end{aligned} \tag{4.12}$$

Thus, $\tau \leq T - r_0$. Letting $X(t) = Y(t)$ for $t \in [\tau, T]$, from (4.7) we see that $Y(t)$ solves (4.10) for $t \in [\tau, T]$ as well (this is why we take in (4.10) the delay term $b(t,X_t)$ rather than $b(t,Y_t)$). In particular, $X_T = Y_T$.

Now let $\mathrm{d}\mathbb{Q} = R\mathrm{d}\mathbb{P}$, where

$$\begin{aligned} R &:= \exp\left[-\int_0^T \langle \Gamma(t), \mathrm{d}B(t)\rangle - \frac{1}{2}\int_0^T |\Gamma(t)|^2\mathrm{d}t\right], \\ \Gamma(t) &:= \frac{\gamma(t)\sigma(t)^{-1}(X(t)-Y(t))}{|X(t)-Y(t)|} + b(t,X_t) - b(t,Y_t). \end{aligned}$$

Then the distribution of Y_T under $\mathbb{Q}$ coincides with that of X_T^η under $\mathbb{P}$. Noting that $\frac{\mathrm{e}^{Kt}(\mathrm{e}^{-2Kt}-\mathrm{e}^{-2K(T-r_0)})}{1-\mathrm{e}^{-2K(T-r_0)}}$ is decreasing in $t \in [0, T-r_0]$, (4.11), (4.12), (4.8), (4.9), and $\tau \leq T - r_0$ imply

$$\begin{aligned} &\left(\int_0^T |\Gamma(t)|^2\mathrm{d}t\right)^{\frac{1}{2}} \leq \left(\frac{1}{\lambda}\int_0^{T-r_0} \gamma(t)^2\mathrm{d}t\right)^{\frac{1}{2}} + \left(\int_0^T (K'\|X_t - Y_t\|_\infty)^2\mathrm{d}t\right)^{\frac{1}{2}} \\ &= |\xi(0)-\eta(0)|\left(\frac{2K}{\lambda(1-\mathrm{e}^{-2K(T-r_0)})}\right)^{\frac{1}{2}} \\ &\quad + K'\left(r_0\|\xi-\eta\|_\infty^2 + |\xi(0)-\eta(0)|^2\int_0^{T-r_0}\frac{\mathrm{e}^{2Kt}(\mathrm{e}^{-2Kt}-\mathrm{e}^{-2K(T-r_0)})^2}{(1-\mathrm{e}^{-2K(T-r_0)})^2}\mathrm{d}t\right)^{\frac{1}{2}} \\ &\leq C(T,\xi,\eta). \end{aligned}$$

So

$$\begin{aligned} \mathbb{E}\{R\log R\} &= \mathbb{E}_\mathbb{Q}\int_0^\tau |\Gamma(t)|^2\mathrm{d}t \leq C(T,\xi,\eta)^2, \\ \mathbb{E}R^{\frac{p}{p-1}} &\leq \left(\mathbb{E}\mathrm{e}^{-\frac{p}{p-1}\int_0^\tau\langle\Gamma(t),\mathrm{d}B(t)\rangle - \frac{p^2}{2(p-1)^2}\int_0^\tau|\Gamma(t)|^2\mathrm{d}t}\right)\mathrm{e}^{\frac{p}{2(p-1)^2}C(T,\xi,\eta)^2} \end{aligned}$$

$$= e^{\frac{p}{2(p-1)^2}C(T,\xi,\eta)^2}.$$

Therefore, the proof is finished by Theorem 1.1.1. □

Theorem 4.2.2 *Let $T > r_0$. Assume that b and a have directional derivatives in the second variable that are bounded uniformly in $t \in [0,T]$. For every $\eta \in \mathscr{C}$ and $t \in [0, T - r_0]$, let $\eta_t \in \mathscr{C}$ be defined by $\eta_t(s) = \eta(t+s)$, $s \in [-r_0, 0]$ for*

$$\eta(s) := 1_{[-r_0,0]}(s)\eta(s) + 1_{(0,T-r_0]}(s)\frac{T - r_0 - s}{T - r_0}\eta(0).$$

Then for every $f \in \mathscr{B}_b(\mathscr{C})$,

$$\nabla_\eta P_T f = \mathbb{E}\int_0^T \langle \sigma(t)^{-1}\Gamma(t), \mathrm{d}B(t)\rangle,$$

where $\Gamma(t) := \frac{\eta(0)1_{[0,T-r_0]}(t)}{T-r_0} + \frac{(T-r_0-t)^+}{T-r_0}\nabla_{\eta(0)}a(t,\cdot)(X(t)) + \nabla_{\eta_t}b(t,\cdot)(X_t)$.

Proof. Let $X(t)$ solve (4.7) for $X_0 = \xi$, and for every $\varepsilon \in (0,1)$ let $X^\varepsilon(t)$ solve the equation

$$\mathrm{d}X^\varepsilon(t) = \Big\{a(t, X(t)) + b(t, X_t) - \frac{\varepsilon 1_{[0,T-r_0)}(t)\eta(0)}{T - r_0}\Big\}\mathrm{d}t + \sigma(t)\mathrm{d}B(t)$$

for $X_0^\varepsilon = \xi + \varepsilon\eta$. Then

$$X^\varepsilon(t) - X(t) = \frac{\varepsilon(T - r_0 - t)}{T - r_0}1_{[0,T-r_0)}(t)\eta(0), \quad t \in [0,T]. \tag{4.13}$$

In particular, $X_T^\xi = X_T$. Let $\mathrm{d}\mathbb{Q}_\varepsilon = R_\varepsilon \mathrm{d}\mathbb{P}$, where

$$R_\varepsilon := \exp\left[\int_0^T \langle \sigma(t)^{-1}\Gamma_\varepsilon(t), \mathrm{d}B(t)\rangle - \frac{1}{2}\int_0^T |\Gamma_\varepsilon(t)|^2 \mathrm{d}t\right],$$

$$\Gamma_\varepsilon(t) := \frac{\varepsilon\eta(0)1_{[0,T-r_0]}(t)}{T - r_0} + a(t, X^\varepsilon(t)) - a(t, X(t)) + b(t, X_t^\varepsilon) - b(t, X_t).$$

Then the distribution of X_T^ξ under $\mathbb{Q}_\varepsilon$ coincides with that of $X_T^{\xi+\varepsilon\eta}$ under $\mathbb{P}$. By (4.13), we see that as $\varepsilon \to 0$, $\frac{R_\varepsilon - 1}{\varepsilon}$ converges in $L^1(\mathbb{P})$ to $\int_0^T \langle \sigma(t)^{-1}\Gamma(t), \mathrm{d}B(t)\rangle$. Then the proof is finished by Theorem 1.1.2. □

4.2.2 Shift Harnack Inequalities and Integration by Parts Formulas

In this subsection, we aim to establish the integration by parts formula and shift Harnack inequality for P_T associated to (4.7). It turns out that we are able to make derivatives or shifts only along directions in the Cameron–Martin space

$$\mathbb{H}^1 := \left\{ h \in \mathscr{C} : \|h\|_{\mathbb{H}^1}^2 := \int_{-r_0}^{0} |h'(t)|^2 \mathrm{d}t < \infty \right\}.$$

Theorem 4.2.3 *Let $T > r_0$ and $\eta \in \mathbb{H}^1$ be fixed. Let $\bar{b}(t,\xi) = a(t,\xi(0)) + b(t,\xi)$. For every $\phi \in \mathscr{B}_b([0,T-r_0])$ such that $\int_0^{T-r_0} \phi(t)\mathrm{d}t = 1$, let*

$$\Phi(t) = 1_{[0,T-r_0]}(t)\phi(t)\eta(-r_0) + 1_{(T-r_0,T]}(t)\eta'(t-T).$$

If

$$\frac{1}{\lambda} := \sup_{t\in[0,T]} \|\sigma(t)^{-1}\|^2 < \infty, \quad \kappa := \sup_{t\in[0,T],\|\xi\|_\infty \le 1} \|\nabla_\xi \bar{b}(t,\cdot)\|_\infty < \infty, \tag{4.14}$$

then:

(1) *For every $f \in \mathscr{B}_b(\mathscr{C})$ with bounded $\nabla_\eta f$,*

$$P_T(\nabla_\eta f) = \mathbb{E}\left(f(X_T) \int_0^T \left\langle \sigma(t)^{-1}(\Phi(t) - \nabla_{\Theta_t}\bar{b}(t,\cdot)(X_t)), \mathrm{d}B(t) \right\rangle \right)$$

holds for

$$\Theta(t) = \int_0^{t\vee 0} \Phi(s)\mathrm{d}s, \quad t \in [-r_0, T].$$

Consequently, for every $\delta > 0$ and positive f,

$$|P_T(\nabla_\eta f)| \le \delta\{P_T(f\log f) - (P_T f)\log P_T f\} + \frac{2(1+\kappa^2 T^2)}{\lambda\delta}\left(\|\eta\|_{\mathbb{H}^1}^2 + \frac{|\eta(-r_0)|^2}{T-r_0}\right) P_T f.$$

(2) *For every nonnegative $f \in \mathscr{B}_b(\mathscr{C})$ and $p > 1$,*

$$(P_T f)^p \le (P_T\{f(\eta+\cdot)\}^p) \exp\left[\frac{2p(1+\kappa^2T^2)}{\lambda(p-1)}\left(\|\eta\|_{\mathbb{H}^1}^2 + \frac{|\eta(-r_0)|^2}{T-r_0}\right)\right].$$

(3) *For every positive $f \in \mathscr{B}_b(\mathscr{C})$,*

$$P_T \log f \le \log P_T\{f(\eta+\cdot)\} + \frac{2(1+\kappa^2T^2)}{\lambda}\left(\|\eta\|_{\mathbb{H}^1}^2 + \frac{|\eta(-r_0)|^2}{T-r_0}\right).$$

Proof. Obviously, the second condition in (4.14) and the boundedness of $\sigma(t)$ for $t \in [0,T]$ imply **(A4.1)** with $u(r) = c$ for some constant $c > 0$. So by Corollary 4.1.2, (4.7) has a unique strong solution. For fixed $\xi \in \mathscr{C}$, let $X(t)$ solve (4.7) for $X_0 = \xi$. For every $\varepsilon \in [0,1]$, let $X^\varepsilon(t)$ solve the equation

$$\mathrm{d}X^\varepsilon(t) = \{\bar{b}(t,X_t) + \varepsilon\Phi(t)\}\mathrm{d}t + \sigma(t)\mathrm{d}B(t), \quad t \ge 0, X_0^\varepsilon = \xi.$$

Then it is easy to see that

$$X_t^\varepsilon = X_t + \varepsilon\Theta_t,\ \ t \in [0,T]. \tag{4.15}$$

In particular, $X_T^\varepsilon = X_T + \varepsilon\eta$. Next, let

$$\begin{aligned} R_\varepsilon = \exp\bigg[& -\int_0^T \Big\langle \sigma(t)^{-1}\{\varepsilon\Phi(t) + \bar{b}(t,X_t) - \bar{b}(t,X_t^\varepsilon)\}, \mathrm{d}B(t)\Big\rangle \\ & -\frac{1}{2}\int_0^T \Big|\sigma(t)^{-1}\{\varepsilon\Phi(t) + \bar{b}(t,X_t) - \bar{b}(t,X_t^\varepsilon)\}\Big|^2 \mathrm{d}t\bigg]. \end{aligned}$$

By Girsanov's theorem, under the changed probability $\mathbb{Q}_\varepsilon := R_\varepsilon \mathbb{P}$, the process

$$B^\varepsilon(t) := B(t) + \int_0^t \sigma(s)^{-1}\big(\varepsilon\Phi(s) + \bar{b}(s,X_s) - \bar{b}(s,X_s^\varepsilon)\big)\mathrm{d}s,\ \ t \in [0,T]$$

is a d-dimensional Brownian motion. So (X_t, X_t^ε) is a coupling by change of measure with changed probability $\mathbb{Q}_\varepsilon$. Then the desired integration by parts formula follows from Theorem 1.1.3, since $R_0 = 1$ and due to (4.15),

$$\frac{\mathrm{d}}{\mathrm{d}\varepsilon}R^\varepsilon\bigg|_{\varepsilon=0} = -\int_0^T \Big\langle \sigma(t)^{-1}(\Phi(t) - \nabla_{\Theta_t}\bar{b}(t,\cdot)(X_t)), \mathrm{d}B(t)\Big\rangle$$

holds in $L^1(\mathbb{P})$. Taking $\phi(t) = \frac{1}{T-r_0}$, we have

$$\begin{aligned} & \int_0^T |\Phi(t)|^2 \mathrm{d}t \le \|\eta\|_{\mathbb{H}}^2 + \frac{|\eta(-r_0)|^2}{T-r_0}, \\ & \|\nabla_{\Theta_t}\bar{b}(t,\cdot)\|_\infty^2 \le \kappa^2\Big(\int_0^T |\Phi(t)|\mathrm{d}t\Big)^2 \le \kappa^2 T\int_0^T |\Phi(t)|^2 \mathrm{d}t. \end{aligned}$$

Then

$$\int_0^T |\Phi(t) - \nabla_{\Theta_t}\bar{b}(t,\cdot)(X_t)|^2 \mathrm{d}t \le 2(1+T^2\kappa^2)\Big(\|\eta\|_{\mathbb{H}}^2 + \frac{|\eta(-r_0)|^2}{T-r_0}\Big). \tag{4.16}$$

So

$$\begin{aligned} & \log\mathbb{E}\exp\bigg[\frac{1}{\delta}\int_0^T \Big\langle \sigma(t)^{-1}\{\Phi(t) - \nabla_{\Theta_t}\bar{b}(t,\cdot)(X_t)\}, \mathrm{d}B(t)\Big\rangle\bigg] \\ & \le \frac{1}{2}\log\mathbb{E}\exp\bigg[\frac{2}{\lambda\delta^2}\int_0^T \big|\Phi(t) - \nabla_{\Theta_t}\bar{b}(t,\cdot)(X_t)\big|^2 \mathrm{d}t\bigg] \\ & \le \frac{2(1+T^2\kappa^2)}{\lambda\delta^2}\Big(\|\eta\|_{\mathbb{H}}^2 + \frac{|\eta(-r_0)|^2}{T-r_0}\Big). \end{aligned}$$

Then the second result in (1) follows from Young's inequality

$$
\begin{aligned}
|P_T(\nabla_\eta f)| \le\ & \delta\{P_T(f\log f) - (P_T f)\log P_T f\} \\
& + \delta \log \mathbb{E}\exp\left[\frac{1}{\delta}\int_0^T \left\langle \sigma(t)^{-1}\{\Gamma(t) - \nabla_{\Theta_t}\bar{b}(t,\cdot)(X_t)\}, \mathrm{d}B(t)\right\rangle\right].
\end{aligned}
$$

Finally, (2) and (3) can be easily derived by applying Theorem 1.1.3 to the above-constructed coupling with $\varepsilon = 1$ and using (4.15) and (4.16). □

4.2.3 Extensions to Semilinear SDPDEs

Consider (4.6) on $\mathbb{H}$ for $\sigma(t,\xi) = \sigma(t)$, i.e.,

$$
\mathrm{d}X(t) = \{AX(t) + a(t,X(t)) + b(t,X_t)\}\mathrm{d}t + \sigma(t)\mathrm{d}W(t), \quad t \in [0,T].
$$

Let P_T be associated to the segment solution, i.e., $P_T f = \mathbb{E}f(X_T)$. According to Theorem 4.1.3 and Remark 3.2.1, we can easily extend Theorems 4.2.1 and 4.2.2 to the present equation. Since in the present case A is normally unbounded, so that κ defined in Theorem 4.2.3 for $\bar{b}(t,\xi) := A\xi(0) + a(t,\xi(0)) + b(t,\xi)$ is infinite, we are unable to extend this result.

Theorem 4.2.4 *Assume* **(A4.2)** *and* **(A4.3)** *instead of* **(A4.1)**, *and assume that* **(A3.3)** *holds. Then all assertions in Theorems 4.2.1 and 4.2.2 hold for $\mathbb{H}$ and $\mathscr{C}(\mathbb{H})$ in place of $\mathbb{R}^d$ and $\mathscr{C}$ respectively.*

4.3 Elliptic Stochastic Functional Partial Differential Equations with Multiplicative Noise

This section is based on [8]. Consider the following equation on $\mathbb{R}^d$:

$$
\mathrm{d}X(t) = \{a(t,X(t)) + b(t,X_t)\}\mathrm{d}t + \sigma(t,X(t))\mathrm{d}B(t), \quad t \in [0,T], \tag{4.17}
$$

for some $T > r_0$. To establish the Harnack inequalities, we shall need the following assumption; see [47, Sect. 3] for the log-Harnack inequality under a weaker assumption in the spirit of **(A4.1)**.

(A4.4) For $T > r_0$, there exist constants $K_1, K_2 \ge 0$, $K_3 > 0$, and $K_4 \in \mathbb{R}$ such that

$$
\left|\sigma(t,\eta(0))^{-1}\{b(t,\xi) - b(t,\eta)\}\right| \le K_1\|\xi - \eta\|_\infty; \tag{4.18}
$$

$$
\left|(\sigma(t,x) - \sigma(t,y))\right| \le K_2(1 \wedge |x-y|); \tag{4.19}
$$

$$
\left|\sigma(t,x)^{-1}\right| \le K_3; \tag{4.20}
$$

$$
\|\sigma(t,x) - \sigma(t,y)\|_{HS}^2 + 2\langle x-y, a(t,x) - a(t,y)\rangle \le K_4|x-y|^2 \tag{4.21}
$$

hold for $t \in [0,T]$, $\xi,\eta \in \mathscr{C}$, and $x,y \in \mathbb{R}^d$.

Obviously, **(A4.4)** implies **(A4.1)**, so that according to Corollary 4.1.2, (4.17) has a unique strong solution. Let P_T be the associated Markov operator.

4.3.1 Log-Harnack Inequality

Theorem 4.3.1 *Assume* **(A4.4)** *for some* $T > r_0$. *Then the log-Harnack inequality*

$$P_T \log f(\eta) \le \log P_T f(\xi) + H(T,\xi,\eta), \quad \xi,\eta \in \mathscr{C},$$

holds for strictly positive $f \in \mathscr{B}_b(\mathscr{C})$ *and*

$$H(T,\xi,\eta) := \inf_{s\in(0,T-r_0]} \left\{ \frac{2K_3^2K_4|\xi(0)-\eta(0)|^2}{1-\mathrm{e}^{-K_4 s}} + 3K_1^2\left\{\frac{r_0}{2} + 4s\big(1+K_2^2K_3^2\big)\right\}\mathrm{e}^{3K_2^2(K_1^2 s+4)s}\|\xi-\eta\|_\infty^2\right\}.$$

The assertion also holds for P_T *associated to* (4.6) *under the further assumptions* **(A3.3)**, **(A4.2)**, *and* **(A4.3)**.

According to Theorem 4.1.3, we need to prove only the first assertion. To this end, we construct a coupling by change of measure as follows. Let $X(s)$ solve (4.17) with $X_0 = \xi$. For fixed $t_0 \in (0, T-r_0]$, let $\gamma \in C^1([0,t_0])$ be such that $\gamma(r) > 0$ for $r \in [0,t_0)$ and $\gamma(t_0) = 0$. Then let $Y(t)$ solve the equation

$$\begin{aligned} \mathrm{d}Y(t) = &\Big\{a(t,Y(t)) + b(t,X_t)\Big\}\mathrm{d}t + \sigma(t,Y(t))\mathrm{d}B(t) \\ &+ \frac{1_{\{t<t_0\}}}{\gamma(t)}\sigma(t,Y(t))\sigma(t,X(t))^{-1}(X(t)-Y(t))\mathrm{d}t, \quad Y_0 = \eta. \end{aligned} \tag{4.22}$$

The key point of our coupling is that $X(t)$ and $Y(t)$ will move together from time t_0 on, so that $X_T = Y_T$. To this end, we add an extra drift term as in (3.52) by taking

$$\gamma(t) = \frac{2-\theta}{K_4}\big(1-\mathrm{e}^{(t-t_0)K_4}\big), \quad t \in [0,t_0),$$

for a parameter $\theta \in (0,2)$. In this case, we have

$$2 + \gamma'(t) - K_4\gamma(t) = \theta, \quad t \in [0,t_0]. \tag{4.23}$$

Moreover, to ensure that these two processes move together after the coupling time (i.e., the first meeting time), they should solve the same equation from that time on. This is why we have to take the delay term in (4.22) using X_t rather than Y_t, as already explained in the proof of Theorem 4.2.1 for the additive noise setting.

Since the additional drift is singular at time t_0, it is clear only that $Y(t)$ is well solved before time t_0. To solve $Y(t)$ for all $t \in [0,T]$, we need to reformulate the equation using a new Brownian motion determined by the Girsanov transform induced by the coupling.

For $t \geq 0$, let

$$\phi(t) = \sigma(t,Y(t))^{-1}\{b(t,Y_t) - b(t,X_t)\} - \frac{1_{\{t<t_0\}}}{\gamma(t)}\sigma(t,X(t))^{-1}(X(t)-Y(t)).$$

From **(A4.4)**, it is easy to see that

$$R(t) := \exp\left[\int_0^t \langle\phi(s), \mathrm{d}B(s)\rangle - \frac{1}{2}\int_0^t |\phi(s)|^2 \mathrm{d}s\right] \tag{4.24}$$

is a martingale for $t \in [0,t_0)$. We shall further prove that:

(i) $R(t_0) := \lim_{t\uparrow t_0} R(t)$ exists and $\{R(t)\}_{t\in[0,t_0]}$ is a uniformly integrable martingale.

Hence **(i)** is confirmed by Girsanov's theorem; under probability $\mathrm{d}\mathbb{Q}_{t_0} := \mathbb{R}(t_0)\mathrm{d}\mathbb{P}$, the process

$$\tilde{B}(t) := B(t) - \int_0^t \langle\phi(s), \mathrm{d}s\rangle \tag{4.25}$$

for $t \in [0,t_0]$ is a d-dimensional Brownian motion; and (4.22) reduces to

$$\mathrm{d}Y(t) = \big\{a(t,Y(t)) + b(t,Y_t)\big\}\mathrm{d}t + \sigma(t,Y(t))\mathrm{d}\tilde{B}(t), \;\; Y_0 = \eta \tag{4.26}$$

for $t \in [0,t_0]$. Therefore, (4.22) has a unique solution $\{Y(t)\}_{t\in[0,t_0]}$ under the probability $\mathbb{Q}_{t_0}$. Moreover, as in Sect. 3.4.1, we have the following:

(ii) The coupling time τ satisfies $\tau := \inf\{t \in [0,t_0] : X(t) = Y(t)\} \leq t_0$, $\mathbb{Q}_{t_0}$-a.s.

Now let $Y(t) = X(t)$ for $t \in [t_0,T]$. Then we will furthermore prove that $\{R(t)\}_{t\in[0,T]}$ given in (4.24) is a uniformly integrable martingale that satisfies the following condition:

(iii) For every $t \in [t_0,T]$,

$$\begin{aligned}\mathbb{E}\big[R(t)\log R(t)\big] \leq\; & \frac{2K_3^2 K_4 |\xi(0) - \eta(0)|^2}{1 - \mathrm{e}^{-K_4 t_0}} \\ & + 3K_1^2\left\{\frac{r_0}{2} + t_0(1 + K_2^2 K_3^2)\right\}\mathrm{e}^{3K_2^2(K_1^2 t_0 + 4)t_0}\|\xi - \eta\|_\infty^2.\end{aligned}$$

By **(i)–(iii)**, we conclude that $\{\tilde{B}(t)\}_{t\in[0,T]}$ given by (4.25) is a d-dimensional Brownian motion under the probability $\mathrm{d}\mathbb{Q}_T := R(T)\mathrm{d}\mathbb{P}$. Moreover, $Y(t)$ solves (4.22) and thus (4.26) up to time T. Thus, $Y_T = X_T$, $\mathbb{Q}_T$-a.s., and the distribution of Y_T under $\mathbb{Q}_T$ coincides with that of X_T^η under $\mathbb{P}$. Therefore, the desired log-Harnack inequality follows from Theorem 1.1.1 and **(iii)** for $t = T$.

Below, we prove assertions **(i)–(iii)**, which imply the desired log-Harnack inequality as explained above.

4.3.1.1 Proofs of (i)

The key result of this subsection is the following.

Proposition 4.3.2 *Assume* **(A4.4)**. *Then for every* $t \in [0,t_0)$,

$$\mathbb{E}\big[R(t)\log R(t)\big] \le \frac{2K_3^2K_4|\xi(0)-\eta(0)|^2}{\theta(2-\theta)(1-\mathrm{e}^{-K_4t_0})} + \frac{12tK_1^2(1+K_2^2K_3^2)\mathrm{e}^{3K_2^2(K_1^2t+4)t}}{\theta^2}\|\xi-\eta\|_\infty^2.$$

Proof. Let $t\in(0,t_0)$ be fixed. Then $\{\tilde B(s)\}_{s\le t}$ is a d-dimensional Brownian motion under the probability $\mathrm{d}\mathbb{Q}_t := R(t)\mathrm{d}\mathbb{P}$, so that (4.18) and (4.20) imply

$$\begin{aligned}
\mathbb{E}[R(t)\log R(t)] &= \mathbb{E}_{\mathbb{Q}_t}\log R(t) = \mathbb{E}_{\mathbb{Q}_t}\left\{\int_0^t\langle\phi(s),\mathrm{d}\tilde B(s)\rangle + \frac12\int_0^t|\phi(s)|^2\mathrm{d}s\right\}\\
&= \frac12\int_0^t\mathbb{E}_{\mathbb{Q}_t}|\phi(s)|^2\mathrm{d}s \qquad (4.27)\\
&\le K_1^2\int_0^t\mathbb{E}_{\mathbb{Q}_t}\|Y_s-X_s\|_\infty^2\mathrm{d}s + K_3^2\int_0^t\frac{1}{\gamma(s)^2}\mathbb{E}_{\mathbb{Q}_t}|X(s)-Y(s)|^2\mathrm{d}s\\
&=: I_1+I_2.
\end{aligned}$$

To estimate I_1 and I_2, let us reformulate (4.17) using the new Brownian motion $\tilde B(s)$:

$$\mathrm{d}X(s) = \big\{a(s,X(s)) + b(s,X_s) + \sigma(s,X(s))\phi(s)\big\}\mathrm{d}s + \sigma(s,X(s))\mathrm{d}\tilde B(s),\quad s\le t.$$

Since

$$\begin{aligned}
&\sigma(s,X(s))\phi(s) + b(s,X_s) - b(s,Y_s)\\
&= \big\{\sigma(s,X(s)) - \sigma(s,Y(s))\big\}\sigma(s,Y(s))^{-1}(b(s,Y_s)-b(s,X_s)) - \frac{X(s)-Y(s)}{\gamma(s)},
\end{aligned}$$

the equation reduces to

$$\begin{aligned}
\mathrm{d}X(s) = \Big\{ &\{\sigma(s,X(s))-\sigma(s,Y(s))\}\sigma(s,Y(s))^{-1}(b(s,Y_s)-b(s,X_s)) \qquad (4.28)\\
&+a(s,X(s)) + b(s,Y_s) - \frac{X(s)-Y(s)}{\gamma(s)}\Big\}\mathrm{d}s + \sigma(s,X(s))\mathrm{d}\tilde B(s),\quad s\le t.
\end{aligned}$$

Combining this with (4.26) and using Itô's formula, we obtain from **(A4.4)** that

$$\begin{aligned}
\mathrm{d}|X(s)-Y(s)|^2 &\le 2\langle X(s)-Y(s),(\sigma(s,X(s))-\sigma(s,Y(s)))\mathrm{d}\tilde B(s)\rangle \qquad (4.29)\\
&+\left\{2K_1K_2\|X_s-Y_s\|_\infty|X(s)-Y(s)| + \left(K_4-\frac{2}{\gamma(s)}\right)|X(s)-Y(s)|^2\right\}\mathrm{d}s,\ s\le t.
\end{aligned}$$

Since it is easy to see that

$$K_4 \le \frac{2}{\gamma(0)} \le \frac{2}{\gamma(s)},$$

it follows that

$$\mathrm{d}|X(s)-Y(s)| \leq \left\langle \frac{X(s)-Y(s)}{|X(s)-Y(s)|}, (\sigma(s,X(s))-\sigma(s,Y(s)))\mathrm{d}\tilde{B}(s)\right\rangle \quad (4.30)$$
$$+K_1K_2\|X_s-Y_s\|_\infty \mathrm{d}s, \ \ s\leq t.$$

Let

$$M(s)=\int_0^s \left\langle \frac{X(r)-Y(r)}{|X(r)-Y(r)|}, (\sigma(r,X(r))-\sigma(r,Y(r)))\mathrm{d}\tilde{B}(r)\right\rangle, \ \ s\leq t,$$

which is a martingale under $\mathbb{Q}_t$. By (4.19) and Doob's inequality, we have

$$\mathbb{E}_{\mathbb{Q}_t}\sup_{r\in[0,s]} M(r)^2 \leq 4K_2^2\int_0^s \mathbb{E}_{\mathbb{Q}_t}\|X_r-Y_r\|_\infty^2 \mathrm{d}r, \ \ s\leq t.$$

Combining this with (4.30), we obtain

$$\mathbb{E}_{\mathbb{Q}_t}\|X_s-Y_s\|_\infty^2 \leq 3\|\xi-\eta\|_\infty^2 + 3K_2^2(K_1^2s+4)\int_0^s \mathbb{E}_{\mathbb{Q}_t}\|X_r-Y_r\|_\infty^2 \mathrm{d}r, \ \ s\leq t.$$

By Gronwall's lemma, this implies that

$$\mathbb{E}_{\mathbb{Q}_t}\|X_s-Y_s\|_\infty^2 \leq 3\|\xi-\eta\|_\infty^2 \mathrm{e}^{3K_2^2(K_1^2s+4)s}, \ \ s\leq t. \quad (4.31)$$

On the other hand, let

$$\mathrm{d}\tilde{M}(s)=\frac{2}{\gamma(s)}\langle X(s)-Y(s), (\sigma(s,X(s))-\sigma(s,Y(s)))\mathrm{d}\tilde{B}(s)\rangle, \ \ s\leq t,$$

which is a martingale under $\mathbb{Q}_t$. It follows from (4.29) and (4.23) that

$$\begin{aligned}
&\mathrm{d}\frac{|X(s)-Y(s)|^2}{\gamma(s)} - \mathrm{d}\tilde{M}(s) \qquad (4.32)\\
&\leq \left\{\frac{2K_1K_2}{\gamma(s)}\|X_s-Y_s\|_\infty|X(s)-Y(s)| + \frac{K_4\gamma(s)-2-\gamma'(s)}{\gamma(s)^2}|X(s)-Y(s)|^2\right\}\mathrm{d}s\\
&\leq \left\{\frac{2K_1K_2}{\gamma(s)}\|X_s-Y_s\|_\infty|X(s)-Y(s)| - \frac{\theta|X(s)-Y(s)|^2}{\gamma(s)^2}\right\}\mathrm{d}s, \ \ s\leq t.
\end{aligned}$$

Combining this with (4.31), we obtain

$$\begin{aligned}
h(t) &:= \int_0^t \frac{\mathbb{E}_{\mathbb{Q}_t}|X(s)-Y(s)|^2}{\gamma(s)^2}\mathrm{d}s\\
&\leq \frac{|\xi(0)-\eta(0)|^2}{\theta\gamma(0)} + \frac{2K_1K_2}{\theta}h(t)^{\frac{1}{2}}\left(\int_0^t \mathbb{E}_{\mathbb{Q}_t}\|X_s-Y_s\|_\infty^2\mathrm{d}s\right)^{\frac{1}{2}}
\end{aligned}$$

$$\leq \frac{|\xi(0)-\eta(0)|^2}{\theta\gamma(0)} + \frac{h(t)}{2} + \frac{6K_1^2K_2^2t}{\theta^2} e^{3K_2^2(K_1^2t+4)t} \|\xi-\eta\|_\infty^2.$$

Therefore,

$$\int_0^t \frac{\mathbb{E}_{\mathbb{Q}_t}|X(s)-Y(s)|^2}{\gamma(s)^2} ds \tag{4.33}$$
$$\leq \frac{2|\xi(0)-\eta(0)|^2}{\theta\gamma(0)} + \frac{12K_1^2K_2^2te^{3K_2^2(K_1^2t+4)t}}{\theta^2} \|\xi-\eta\|_\infty^2.$$

Substituting this and (4.31) into (4.27), we complete the proof. □

Now, according to Proposition 4.3.2, $\{R(t)\}_{t\in[0,t_0)}$ is a uniformly integrable continuous martingale. So by the martingale convergence theorem,

$$R(t_0) = \lim_{t\uparrow t_0} R(t) \tag{4.34}$$

exists and $\{R(t)\}_{t\in[0,t_0]}$ is again a uniformly integrable martingale.

4.3.1.2 Proof of (ii)

Since (4.33) holds for $\mathbb{Q}_{t_0}$ in place of $\mathbb{Q}_t$, letting $t \uparrow t_0$, we conclude that

$$\mathbb{E}_{\mathbb{Q}_{t_0}} \int_0^{t_0} \frac{|X(t)-Y(t)|^2}{\gamma(t)^2} dt < \infty. \tag{4.35}$$

This implies that $\tau \leq t_0$, $\mathbb{Q}_{t_0}$-a.s. Indeed, since $X(t)$ and $Y(t)$ are continuous in $t \in [0,t_0]$, $\mathbb{Q}_{t_0}$-a.s., there exists $\Omega_0 \subset \Omega$ with $\mathbb{Q}_{t_0}(\Omega_0) = 1$ such that for every $\omega \in \Omega_0$, $X(t)(\omega)$ and $Y(t)(\omega)$ are continuous in t. If $\omega \in \Omega_0$ is such that $\tau(\omega) > t_0$, then

$$\inf_{t\in[0,t_0]} |X(t)-Y(t)|(\omega) > 0,$$

so that

$$\int_0^{t_0} \frac{|X(t)-Y(t)|^2(\omega)}{\gamma(t)^2} dt \geq \inf_{t\in[0,t_0]} |X(t)-Y(t)|(\omega) \int_0^{t_0} \frac{dt}{\gamma(t)^2} = \infty.$$

This means that

$$\mathbb{Q}_{t_0}(\tau > t_0) \leq \mathbb{Q}_{t_0}\left(\int_0^{t_0} \frac{|X(t)-Y(t)|^2}{\gamma(t)^2} dt = \infty\right),$$

which equals zero according to (4.35).

4.3.1.3 Proof of (iii)

Let $T' \in [t_0, T]$ be fixed, and let $\tau_n = T \wedge \inf\{t \in [t_0, T] : \int_0^t |\phi(s)|^2 ds \geq n\}$. Then due to **(i)**, $\{R(t)\}_{t\in[0,\tau_n\wedge T']}$ is a martingale, and by Girsanov's theorem, $\{\tilde{B}(t)\}_{t\in[0,\tau_n\wedge T']}$ is a Brownian motion under $d\tilde{\mathbb{Q}}_n := R(\tau_n \wedge T')d\mathbb{P}$. Thus,

$$\begin{aligned}\mathbb{E}[R(\tau_n\wedge T')\log R(\tau_n\wedge T')] &= \frac{1}{2}\mathbb{E}_{\tilde{\mathbb{Q}}_n}\int_0^{\tau_n\wedge T'}|\phi(t)|^2dt\\ &= \frac{1}{2}\mathbb{E}_{\mathbb{Q}_{t_0}}\int_0^{t_0}|\phi(t)|^2dt + \frac{1}{2}\mathbb{E}_{\tilde{\mathbb{Q}}_n}\int_{t_0}^{\tau_n\wedge T'}|\phi(t)|^2dt\\ &= \mathbb{E}[R(t_0)\log R(t_0)] + \frac{1}{2}\mathbb{E}_{\tilde{\mathbb{Q}}_n}\int_{t_0}^{T}|\phi(t)|^2dt.\end{aligned} \tag{4.36}$$

Since $X(t) = Y(t)$ for $t \geq t_0$, it follows from (4.18) that

$$\int_{t_0}^T |\phi(t)|^2dt \leq K_1^2\int_{t_0}^T \|X_t - Y_t\|_\infty^2 dt \leq K_1^2 r_0\|Y_{t_0} - X_{t_0}\|_\infty^2.$$

Noting that

$$\mathbb{E}_{\tilde{\mathbb{Q}}_n}\|Y_{t_0} - X_{t_0}\|_\infty^2 = \mathbb{E}_{\mathbb{Q}_{t_0}}\|Y_{t_0} - X_{t_0}\|_\infty^2,$$

combining this with (4.31), which also holds for $t = s = t_0$ due to (4.34) and Fatou's lemma, we arrive at

$$\frac{1}{2}\mathbb{E}_{\tilde{\mathbb{Q}}_n}\int_{t_0}^T|\phi(t)|^2dt \leq \frac{3K_1^2r_0}{2}e^{3K_2^2(K_1^2t_0+4)t_0}\|\xi-\eta\|_\infty^2.$$

Substituting this into (4.36) and noting that (4.34) and Proposition 4.3.2 with $\theta = 1$ imply

$$\begin{aligned}&\mathbb{E}[R(t_0)\log R(t_0)]\\ &\leq \frac{2K_3^2K_4|\xi(0)-\eta(0)|^2}{1-e^{-K_4t_0}} + 12t_0K_1^2(1+K_2^2K_3^2)e^{3K_2^2(K_1^2t_0+4)t_0}\|\xi-\eta\|_\infty^2,\end{aligned} \tag{4.37}$$

we prove **(iii)**.

4.3.2 Harnack Inequality with Power

In this subsection we aim to establish the Harnack inequality with a power $p > 1$,

$$P_Tf(\eta) \leq \{P_Tf^p(\xi)\}^{\frac{1}{p}}\exp[\Phi_p(T,\xi,\eta)],\ \ f\geq 0,\ T > r_0,\ \xi,\eta\in\mathscr{C}, \tag{4.38}$$

for some positive function Φ_p on $(r_0,\infty)\times\mathscr{C}^2$. As in Theorem 3.4.1 for the case without delay, in the present setting we will need to assume that $p > (1+K_2K_3)^2$.

Let

$$\lambda_p = \frac{1}{2(\sqrt{p}-1)^2}.$$

Then the set

$$\Lambda_p := \left\{\varepsilon \in (0,1) : \frac{(1-\varepsilon)^4}{2(1+\varepsilon)^3 K_2^2 K_3^2} \geq \lambda_p\right\}$$

is nonempty. Let

$$\begin{aligned} W_\varepsilon(\lambda) = \ & \max\left\{\frac{8(1+\varepsilon)r_0K_1^3K_2\lambda\{4(1+\varepsilon)r_0K_1K_2\lambda+\varepsilon\}}{\varepsilon^2},\right. \\ & \left.\frac{2K_1^2(1+\varepsilon)^2\lambda}{\varepsilon^2}, \quad \frac{(1+\varepsilon)^3K_1^2K_2^2K_3^2\lambda}{\varepsilon^2(1-\varepsilon)^3}\right\}, \\ s_\varepsilon(\lambda) = \ & \frac{\sqrt{K_1^2+2W_\varepsilon(\lambda)}-K_1}{4W_\varepsilon(\lambda)K_2}, \quad \varepsilon \in (0,1), \lambda > 0. \end{aligned}$$

Theorem 4.3.3 *Assume* **(A4.4)**. *For every $p > (1+K_2K_3)^2$, the Harnack inequality* (4.38) *holds for*

$$\begin{aligned} \Phi_p(T,\xi,\eta) := \frac{\sqrt{p}-1}{\sqrt{p}} & \inf_{\varepsilon\in\Lambda_p, s\in(0,s_\varepsilon(\lambda_p)\wedge(T-r_0)]}\left\{\frac{\varepsilon}{2(1+\varepsilon)} + \frac{16K_2^2s^2W_\varepsilon(\lambda_p)}{1-4K_1K_2s}\right. \\ & \left. + \frac{\lambda_p(1+\varepsilon)^2K_3^2K_4|\xi(0)-\eta(0)|^2}{2\varepsilon(1-\varepsilon)^2(1+2\varepsilon)(1-\mathrm{e}^{-K_4s})} + (K_1^2r_0\lambda_p + 2sW_\varepsilon(\lambda_p))\|\xi-\eta\|_\infty^2\right\}. \end{aligned}$$

Consequently, there exists a decreasing function $C: ((1+K_2K_3)^2, \infty) \to (0,\infty)$ such that (4.38) *holds for*

$$\Phi_p(T,\xi,\eta) = C(p)\left\{1 + \frac{|\xi(0)-\eta(0)|^2}{T-r_0} + \|\xi-\eta\|_\infty^2\right\}.$$

The assertion also holds for P_T associated to (4.6) *under the further assumptions* **(A3.3)**, **(A4.2)**, *and* **(A4.3)**.

Proof. According to Theorem 4.1.3, we need to prove only the first assertion.

(a) We first observe that the second assertion is a consequence of the first. Indeed, for $q > (1+K_2K_3)^2$, we take $(\varepsilon, s) = (\varepsilon_q, s_q(T))$ for a fixed $\varepsilon_q \in \Lambda_q$ and $s_q(T) := s_{\varepsilon_q}(\lambda_q) \wedge (T-r_0)$. By the definition of Φ_q, there exist two positive constants $c_1(q)$ and $c_2(q)$ such that

$$\begin{aligned} \Phi_q(T,\xi,\eta) & \leq c_1(q)\left(1 + \|\xi-\eta\|_\infty^2 + \frac{|\xi(0)-\eta(0)|^2}{c_2(q)\wedge(T-r_0)}\right) \\ & \leq \frac{c_1(q)(1+c_2(q))}{c_2(q)}\left(1+\|\xi-\eta\|_\infty^2 + \frac{|\xi(0)-\eta(0)|^2}{T-r_0}\right), \quad \xi,\eta \in \mathscr{C}. \end{aligned}$$

So for $p>(1+K_2K_3)^2$ and $q\in((1+K_2K_3)^2,p]$, by the first assertion and using Jensen's inequality, we obtain

$$\begin{aligned} P_Tf(\eta) &\le (P_Tf^q)^{\frac{1}{q}}(\xi)\exp\left[\frac{c_1(q)(1+c_2(q))}{c_2(q)}\left(1+\|\xi-\eta\|_\infty^2+\frac{|\xi(0)-\eta(0)|^2}{T-r_0}\right)\right]\\ &\le (P_Tf^p)^{\frac{1}{p}}(\xi)\exp\left[\frac{c_1(q)(1+c_2(q))}{c_2(q)}\left(1+\|\xi-\eta\|_\infty^2+\frac{|\xi(0)-\eta(0)|^2}{T-r_0}\right)\right]. \end{aligned}$$

Therefore, the second assertion holds for

$$C(p)=\inf_{q\in((1+K_2K_3)^2,p]}\frac{c_1(q)(1+c_2(q))}{c_2(q)},$$

which is decreasing in p.

(b) To prove the first assertion, let us fix $\varepsilon\in\Lambda_p$ and $t_0\in(0,s_\varepsilon(\lambda_p)\wedge(T-r_0)]$. We shall make use of the coupling constructed in Sect. 4.3.1 for $\theta=2(1-\varepsilon)$. Since $t_0\le T-r_0$ and $X(t)=Y(t)$ for $t\ge t_0$, we have $X_T=Y_T$ and

$$P_Tf(\eta)=\mathbb{E}[R_Tf(Y_T)]=\mathbb{E}[R_Tf(X_T)]\le(P_Tf^p(\xi))^{\frac{1}{p}}(\mathbb{E}R_T^{\frac{p}{p-1}})^{\frac{p-1}{p}}. \tag{4.39}$$

By the definition of $R(T)$ and $\mathbb{Q}_T$, we obtain

$$\begin{aligned} \mathbb{E}R_T^{\frac{p}{p-1}} &= \mathbb{E}_{\mathbb{Q}_T}R_T^{\frac{1}{p-1}}\\ &=\mathbb{E}_{\mathbb{Q}_T}\exp\left[\frac{1}{p-1}\int_0^T\langle\phi(t),\mathrm{d}\tilde B(t)\rangle+\frac{1}{2(p-1)}\int_0^T|\phi(t)|^2\mathrm{d}t\right]\\ &=\mathbb{E}_{\mathbb{Q}_T}\exp\left[\frac{1}{p-1}\int_0^T\langle\phi(t),\mathrm{d}\tilde B(t)\rangle-\frac{\sqrt{p}+1}{2(p-1)^2}\int_0^T|\phi(t)|^2\mathrm{d}t\right.\\ &\qquad\left.+\frac{p+\sqrt{p}}{2(p-1)^2}\int_0^T|\phi(t)|^2\mathrm{d}t\right]\\ &\le\left(\mathbb{E}_{\mathbb{Q}_T}\exp\left[\frac{\sqrt{p}+1}{p-1}\int_0^T\langle\phi(t),\mathrm{d}\tilde B(t)\rangle-\frac{(\sqrt{p}+1)^2}{2(p-1)^2}\int_0^T|\phi(t)|^2\mathrm{d}t\right]\right)^{\frac{1}{1+\sqrt{p}}}\\ &\quad\times\left(\mathbb{E}_{\mathbb{Q}_T}\exp\left[\frac{(\sqrt{p}+1)(p+\sqrt{p})}{2(p-1)^2\sqrt{p}}\int_0^T|\phi(t)|^2\mathrm{d}t\right]\right)^{\frac{\sqrt{p}}{\sqrt{p}+1}}\\ &=\left(\mathbb{E}_{\mathbb{Q}_T}\exp\left[\lambda_p\int_0^T|\phi(t)|^2\mathrm{d}t\right]\right)^{\frac{\sqrt{p}}{\sqrt{p}+1}}. \end{aligned}$$

Combining this with (4.39), we obtain

$$P_Tf(\eta)\le(P_Tf^p)^{\frac{1}{p}}(\xi)\left(\mathbb{E}_{\mathbb{Q}_T}\exp\left[\lambda_p\int_0^T|\phi(t)|^2\mathrm{d}t\right]\right)^{\frac{\sqrt{p}-1}{\sqrt{p}}}.$$

Therefore, to prove the first assertion, it suffices to show that

$$\begin{aligned}
&\mathbb{E}_{\mathbb{Q}_T} \exp\left[\lambda_p \int_0^T |\phi(t)|^2 \mathrm{d}t\right] \\
&\leq \exp\left[\frac{\varepsilon}{2(1+\varepsilon)} + \frac{\lambda_p(1+\varepsilon)^2 K_3^2 K_4 |\xi(0)-\eta(0)|^2}{2\varepsilon(1-\varepsilon)^2(1+2\varepsilon)(1-\mathrm{e}^{-K_4 s})}\right. \\
&\qquad \left. + \frac{16K_2^2 s^2 W_\varepsilon(\lambda_p)}{1-4K_1K_2 s} + (K_1^2 r_0 \lambda_p + 2sW_\varepsilon(\lambda_p))\|\xi-\eta\|_\infty^2\right].
\end{aligned} \tag{4.40}$$

Since $X(t) = Y(t)$ for $t \geq t_0$, it is easy to see from the definition of $\phi(t)$, (4.18), and (4.20) that

$$\begin{aligned}
\int_0^T |\phi(t)|^2 \mathrm{d}t \leq & \int_0^{t_0} \left\{\frac{K_1^2(1+\varepsilon)}{\varepsilon}\|X_t - Y_t\|_\infty^2 + \frac{K_3^2(1+\varepsilon)|X(t)-Y(t)|^2}{\gamma(t)^2}\right\}\mathrm{d}t \\
& + K_1^2 r_0 \|X_{t_0} - Y_{t_0}\|_\infty^2.
\end{aligned}$$

By this and Hölder's inequality, we obtain

$$\begin{aligned}
&\mathbb{E}_{\mathbb{Q}_T} \exp\left[\lambda_p \int_0^T |\phi(t)|^2 \mathrm{d}t\right] \\
&\leq \left(\mathbb{E}_{\mathbb{Q}_T} \exp\left[\lambda_p K_3^2(1+\varepsilon)^2 \int_0^{t_0} \frac{|X(t)-Y(t)|^2}{\gamma(t)^2}\mathrm{d}t\right]\right)^{\frac{1}{1+\varepsilon}} \\
&\quad \times \left(\mathbb{E}_{\mathbb{Q}_T} \exp\left[\frac{2K_1^2(1+\varepsilon)^2\lambda_p}{\varepsilon^2}\int_0^{t_0} \|X_t - Y_t\|_\infty^2 \mathrm{d}t\right]\right)^{\frac{\varepsilon}{2+2\varepsilon}} \\
&\quad \times \left(\mathbb{E}_{\mathbb{Q}_T} \exp\left[\frac{2K_1^2 r_0(1+\varepsilon)\lambda_p}{\varepsilon}\|X_{t_0} - Y_{t_0}\|_\infty^2\right]\right)^{\frac{\varepsilon}{2+2\varepsilon}}.
\end{aligned} \tag{4.41}$$

Since $\varepsilon \in \Lambda_p$ implies that

$$\lambda_p K_3^2(1+\varepsilon)^2 \leq \frac{(1-\varepsilon)^4}{2(1+\varepsilon)K_2^2},$$

it follows from Lemma 4.3.4 below that

$$\begin{aligned}
&\mathbb{E}_{\mathbb{Q}_T} \exp\left[\lambda_p K_3^2(1+\varepsilon)^2 \int_0^{t_0} \frac{|X(t)-Y(t)|^2}{\gamma(t)^2}\mathrm{d}t\right] \\
&\leq \exp\left[\frac{\lambda_p K_3^2(1+\varepsilon)^3|\xi(0)-\eta(0)|^2}{(1+2\varepsilon)(1-\varepsilon)^2\gamma(0)}\right] \\
&\quad \times \left(\mathbb{E}_{\mathbb{Q}_T} \exp\left[\frac{K_1^2K_2^2K_3^2\lambda_p(1+\varepsilon)^3}{\varepsilon^2(1-\varepsilon)^3}\int_0^{t_0}\|X_t - Y_t\|_\infty^2 \mathrm{d}t\right]\right)^{\frac{\varepsilon}{1+2\varepsilon}}.
\end{aligned} \tag{4.42}$$

Moreover, according to Lemma 4.3.5 below,

$$\mathbb{E}_{\mathbb{Q}_T} \exp\left[\frac{2K_1^2 r_0(1+\varepsilon)\lambda_p}{\varepsilon}\|X_{t_0} - Y_{t_0}\|_\infty^2\right] \leq \exp\left[1 + \frac{2K_1^2 r_0(1+\varepsilon)\lambda_p}{\varepsilon}\|\xi-\eta\|_\infty^2\right]$$

$$\times\left(\mathbb{E}_{\mathbb{Q}_T}\exp\left[\frac{8K_1^3K_2r_0(1+\varepsilon)\lambda_p(4K_2K_1r_0(1+\varepsilon)\lambda_p+\varepsilon)}{\varepsilon^2}\int_0^{t_0}\|X_t-Y_t\|_\infty^2\mathrm{d}t\right]\right)^{\frac{1}{2}}.$$

Substituting this and (4.42) into (4.41) and using the definition of $W_\varepsilon(\lambda_p)$, we conclude that

$$\mathbb{E}_{\mathbb{Q}_T}\exp\left[\lambda_p\int_0^T|\phi(t)|^2\mathrm{d}t\right]\le\mathbb{E}_{\mathbb{Q}_T}\exp\left[W_\varepsilon(\lambda_p)\int_0^{t_0}\|X_t-Y_t\|_\infty^2\mathrm{d}t\right] \tag{4.43}$$
$$\times\exp\left[\frac{\lambda_pK_3^2(1+\varepsilon)^2|\xi(0)-\eta(0|^2}{(1+2\varepsilon)(1-\varepsilon)^2\gamma(0)}+\frac{\varepsilon}{2(1+\varepsilon)}+K_1^2r_0\lambda_p\|\xi-\eta\|_\infty^2\right].$$

Since $t_0\le s_\varepsilon(\lambda_p)$, we have

$$W_\varepsilon(\lambda_p)\le\frac{1-4K_1K_2t_0}{8K_2^2t_0^2}.$$

So combining (4.43) with Lemma 4.3.6 below and noting that for $\theta=2(1-\varepsilon)$, we have

$$\gamma(0)=\frac{2\varepsilon}{K_4}(1-\mathrm{e}^{-K_4t_0}),$$

we prove (4.40). □

Lemma 4.3.4 *For every positive* $\lambda\le\frac{(1-\varepsilon)^4}{2K_2^2(1+\varepsilon)}$ *and* $s\in[0,t_0]$,

$$\mathbb{E}_{\mathbb{Q}_T}\exp\left[\lambda\int_0^s\frac{|X(t)-Y(t)|^2}{\gamma(t)^2}\mathrm{d}t\right]$$
$$\le\exp\left[\frac{\lambda(1+\varepsilon)|\xi(0)-\eta(0)|^2}{(1+2\varepsilon)(1-\varepsilon)^2\gamma(0)}\right]\left(\mathbb{E}_{\mathbb{Q}_T}\exp\left[\frac{K_1^2K_2^2(1+\varepsilon)\lambda}{\varepsilon^2(1-\varepsilon)^3}\int_0^s\|X_t-Y_t\|_\infty^2\mathrm{d}t\right]\right)^{\frac{\varepsilon}{1+2\varepsilon}}.$$

Proof. Since $\theta=2(1-\varepsilon)$ and

$$\frac{K_1K_2}{\gamma(t)}\|X_t-Y_t\|_\infty|X(t)-Y(t)|\le\frac{K_1^2K_2^2}{4\theta\varepsilon}\|X_t-Y_t\|_\infty^2+\theta\varepsilon\frac{|X(t)-Y(t)|^2}{\gamma(t)^2},$$

it follows from (4.32) that

$$0\le\tilde{M}(s)+\frac{|\xi(0)-\eta(0)|^2}{\gamma(0)}+\int_0^s\left\{\frac{K_1^2K_2^2\|X_t-Y_t\|_\infty^2}{\varepsilon(1-\varepsilon)}-\frac{2(1-\varepsilon)^2|X(t)-Y(t)|^2}{\gamma(t)^2}\right\}\mathrm{d}t.$$

Combining this with (4.19) and the fact that

$$\mathbb{E}_{\mathbb{Q}_T}\mathrm{e}^{N(s)+L}\le\left(\mathbb{E}_{\mathbb{Q}_T}\mathrm{e}^{2\langle N\rangle(s)+2L}\right)^{\frac{1}{2}} \tag{4.44}$$

holds for a $\mathbb{Q}_T$-martingale N and a random variable L, we obtain

$$\begin{aligned}
&\mathbb{E}_{\mathbb{Q}_T} \exp\left[\lambda \int_0^s \frac{|X(t)-Y(t)|^2}{\gamma(t)^2}\mathrm{d}t - \frac{\lambda|\xi(0)-\eta(0)|^2}{2\gamma(0)(1-\varepsilon)^2}\right] \\
&\le \mathbb{E}_{\mathbb{Q}_T} \exp\left[\frac{\lambda}{2(1-\varepsilon)^2}\tilde{M}(s) + \frac{K_1^2K_2^2\lambda}{2\varepsilon(1-\varepsilon)^3}\int_0^s \|X_t - Y_t\|_\infty^2 \mathrm{d}t\right] \\
&\le \left(\mathbb{E}_{\mathbb{Q}_T} \exp\left[\frac{2K_2^2\lambda^2}{(1-\varepsilon)^4}\int_0^s \frac{|X(t)-Y(t)|^2}{\gamma(t)^2}\mathrm{d}t + \frac{K_1^2K_2^2\lambda}{\varepsilon(1-\varepsilon)^3}\int_0^s \|X_t - Y_t\|_\infty^2 \mathrm{d}t\right]\right)^{\frac{1}{2}} \\
&\le \left(\mathbb{E}_{\mathbb{Q}_T} \exp\left[\frac{2K_2^2(1+\varepsilon)\lambda^2}{(1-\varepsilon)^4}\int_0^s \frac{|X(t)-Y(t)|^2}{\gamma(t)^2}\mathrm{d}t\right]\right)^{\frac{1}{2+2\varepsilon}} \\
&\quad \times \left(\mathbb{E}_{\mathbb{Q}_T} \exp\left[\frac{K_1^2K_2^2(1+\varepsilon)\lambda}{\varepsilon^2(1-\varepsilon)^3}\int_0^s \|X_t - Y_t\|_\infty^2 \mathrm{d}t\right]\right)^{\frac{\varepsilon}{2+2\varepsilon}}.
\end{aligned}$$

Since

$$\frac{2K_2^2(1+\varepsilon)\lambda^2}{(1-\varepsilon)^4} \le \lambda$$

and up to an approximation we may assume that

$$\mathbb{E}_{\mathbb{Q}_T} \exp\left[\lambda \int_0^s \frac{|X(t)-Y(t)|^2}{\gamma(t)^2}\mathrm{d}t\right] < \infty,$$

this implies the desired inequality. □

Lemma 4.3.5 *For every $\lambda > 0$ and $s \in [0, t_0]$,*

$$\mathbb{E}_{\mathbb{Q}_T} \mathrm{e}^{\lambda\|X_s - Y_s\|_\infty^2} \le \mathrm{e}^{1+\lambda\|\xi-\eta\|_\infty^2}\left(\mathbb{E}_{\mathbb{Q}_T} \exp\left[4\lambda K_2(2\lambda K_2 + K_1)\int_0^s \|X_t - Y_t\|_\infty^2 \mathrm{d}t\right]\right)^{\frac{1}{2}}.$$

Proof. Let

$$N(t) = 2\int_0^t \langle X(r) - Y(r), (\sigma(r, X(r)) - \sigma(r, Y(r)))\mathrm{d}\tilde{B}(r)\rangle, \quad r \le s,$$

which is a $\mathbb{Q}_T$-martingale. By (4.29) and noting that $K_4 \le \frac{2}{\gamma(r)}$, we obtain

$$\begin{aligned}
\|X_t - Y_t\|_\infty^2 &\le \left\{\sup_{r\in[0,t]} |X(r) - Y(r)|^2\right\} \vee \|\xi - \eta\|_\infty^2 \\
&\le \|\xi - \eta\|_\infty^2 + \sup_{r\in[0,t]}\left\{N(r) + 2K_1K_2\int_0^r \|X_u - Y_u\|_\infty^2 \mathrm{d}u\right\}.
\end{aligned}$$

Combining this with (4.44) and noting that Doob's inequality implies

$$\begin{aligned}
&\mathbb{E}_{\mathbb{Q}_T} \sup_{r\in[0,t]} \mathrm{e}^{M(r)} = \lim_{p\to\infty} \mathbb{E}_{\mathbb{Q}_T}\left(\sup_{r\in[0,t]} \mathrm{e}^{\frac{M(r)}{p}}\right)^p \\
&\le \lim_{p\to\infty}\left(\frac{p}{p-1}\right)^p \mathbb{E}_{\mathbb{Q}_T}(\mathrm{e}^{\frac{M(t)}{p}})^p = \mathrm{e}\mathbb{E}_{\mathbb{Q}_T}\mathrm{e}^{M(t)}
\end{aligned}$$

for a $\mathbb{Q}_T$-submartingale $M(r)$, we arrive at

$$\begin{aligned}
&\mathbb{E}_{\mathbb{Q}_T} e^{\lambda\|X_s - Y_s\|_\infty^2 - \lambda\|\xi-\eta\|_\infty^2} \le \mathbb{E}_{\mathbb{Q}_T} \sup_{t\in[0,s]} \exp\left[\lambda N(t) + 2\lambda K_1 K_2 \int_0^t \|X_r - Y_r\|_\infty^2 \mathrm{d}r\right] \\
&\le e\,\mathbb{E}_{\mathbb{Q}_T} \exp\left[\lambda N(s) + 2\lambda K_1 K_2 \int_0^s \|X_t - Y_t\|_\infty^2 \mathrm{d}t\right] \\
&\le e\left(\mathbb{E}_{\mathbb{Q}_T} \exp\left[2\lambda^2 \langle N\rangle(s) + 4\lambda K_1 K_2 \int_0^s \|X_t - Y_t\|_\infty^2 \mathrm{d}t\right]\right)^{\frac{1}{2}} \\
&\le e\left(\mathbb{E}_{\mathbb{Q}_T} \exp\left[(8K_2^2\lambda^2 + 4\lambda K_1 K_2) \int_0^s \|X_t - Y_t\|_\infty^2 \mathrm{d}t\right]\right)^{\frac{1}{2}}.
\end{aligned}$$

□

Lemma 4.3.6 *For every $s \in (0, t_0]$ and positive $\lambda \le \frac{1-4K_1K_2 s}{8K_2^2 s^2}$,*

$$\mathbb{E}_{\mathbb{Q}_T} \exp\left[\lambda \int_0^s \|X_t - Y_t\|_\infty^2 \mathrm{d}t\right] \le \exp\left[\frac{16K_2^2 s^2 \lambda}{1-4K_1K_2 s} + 2s\lambda\|\xi-\eta\|_\infty^2\right].$$

Proof. Let

$$\lambda_0 = \frac{1-4K_1K_2 s}{8K_2^2 s^2},$$

which is positive, since $s \in (0, s_\varepsilon(\lambda_p)]$. It is easy to see that

$$4K_2 s\lambda_0(2K_2 s\lambda_0 + K_1) = \lambda_0.$$

So it follows from Jensen's inequality and Lemma 4.3.5 that

$$\begin{aligned}
&\mathbb{E}_{\mathbb{Q}_T} \exp\left[\lambda_0 \int_0^s \|X_t - Y_t\|_\infty^2 \mathrm{d}t\right] \le \frac{1}{s}\int_0^s \mathbb{E}_{\mathbb{Q}_T} e^{\lambda_0 s\|X_t - Y_t\|_\infty^2} \mathrm{d}t \qquad (4.45)\\
&\le e^{1+\lambda_0 s\|\xi-\eta\|_\infty^2}\left(\mathbb{E}_{\mathbb{Q}_T} \exp\left[4\lambda_0 K_2 s(2\lambda_0 K_2 s + K_1) \int_0^s \|X_t - Y_t\|_\infty^2 \mathrm{d}t\right]\right)^{\frac{1}{2}} \\
&= e^{1+\lambda_0 s\|\xi-\eta\|_\infty^2}\left(\mathbb{E}_{\mathbb{Q}_T} \exp\left[\lambda_0 \int_0^s \|X_t - Y_t\|_\infty^2 \mathrm{d}t\right]\right)^{\frac{1}{2}}.
\end{aligned}$$

Up to an approximation argument we may assume that

$$\mathbb{E}_{\mathbb{Q}_T} \exp\left[\lambda_0 \int_0^s \|X_t - Y_t\|_\infty^2 \mathrm{d}t\right] < \infty,$$

so that this implies

$$\mathbb{E}_{\mathbb{Q}_T} \exp\left[\lambda_0 \int_0^s \|X_t - Y_t\|_\infty^2 \mathrm{d}t\right] \le e^{2+2\lambda_0 s\|\xi-\eta\|_\infty^2}.$$

Therefore, by Jensen's inequality, for every $\lambda \in [0,\lambda_0]$,

$$\begin{aligned}\mathbb{E}_{\mathbb{Q}_T} \exp\left[\lambda \int_0^s \|X_t - Y_t\|_\infty^2 \mathrm{d}t\right] &\le \left(\mathbb{E}_{\mathbb{Q}_T} \exp\left[\lambda_0 \int_0^s \|X_t - Y_t\|_\infty^2 \mathrm{d}t\right]\right)^{\frac{\lambda}{\lambda_0}} \\ &\le \exp\left[\frac{2\lambda}{\lambda_0} + 2\lambda s \|\xi - \eta\|_\infty^2\right].\end{aligned}$$

□

4.3.3 Bismut Formulas for Semilinear SDPDEs

Consider the following semilinear SDPDE on a separable Hilbert space $\mathbb{H}$:

$$\mathrm{d}X(t) = \{AX(t) + b(t,X_t)\}\mathrm{d}t + \sigma(t,X(t))\mathrm{d}W(t) \tag{4.46}$$

under the assumption

(A4.5) For some $T > r_0$:

(i) $(A,\mathscr{D}(A))$ is a linear operator on $\mathbb{H}$ generating a contractive C_0-semigroup $(S(t))_{t\ge 0}$.
(ii) $b : [0,T] \times \mathscr{C}(\mathbb{H}) \to H$ is Gâteaux differentiable such that $\nabla b(t,\cdot) : \mathscr{C}(\mathbb{H}) \times \mathscr{C}(\mathbb{H}) \to \mathbb{H}$ is bounded uniformly in $t \in [0,T]$ on $\mathscr{C}(\mathbb{H}) \times \mathscr{C}(\mathbb{H})$ and uniformly continuous on bounded sets.
(iii) $\sigma : [0,T] \times \mathbb{H} \to \mathscr{L}_b(\mathbb{H})$ is bounded and Gâteaux differentiable such that $\nabla\sigma(t,\cdot) : H \times H \to \mathscr{L}_{HS}(\mathbb{H})$ is bounded uniformly in $t \in [0,T]$ on $\mathbb{H} \times \mathbb{H}$ and uniformly continuous on bounded sets, and $\sigma(t,x)$ is invertible for every $x \in \mathbb{H}$ and $t \in [0,T]$.
(iv) $\int_0^t s^{-2\varepsilon} \|S(s)\|_{HS}^2 \mathrm{d}s < \infty$ holds for some constant $\varepsilon \in (0,1)$ and all $t \in [0,T]$.

It is easy to see that **(A4.5)** implies **(A4.2)** and **(A4.3)** for all $\varepsilon > 0$ as well as the condition (3.6) for some $\alpha \in (0,1)$. Then by Theorem 4.1.3, for every $\xi \in \mathscr{C}(\mathbb{H})$, (4.46) has a unique mild solution with $X_0 = \xi$, which is continuous and satisfies

$$\mathbb{E} \sup_{s\in[0,T]} \|X_s\|_\infty^p < \infty, \quad p \ge 1. \tag{4.47}$$

Let P_T be the semigroup for the segment mild solution.

Theorem 4.3.7 *Assume that* **(A4.5)**. *Let $u : [0,\infty) \to \mathbb{R}$ be C^1 such that $u(t) > 0$ for $t \in [0,T-r_0)$, $u(t) = 0$ for $t \ge T - r_0$, and*

$$\theta_p := \inf_{t\in[0,T-r_0]} \{p + (p-1)u'(t)\} > 0$$

holds for some $p > 1$. Then for every $\xi,\eta \in \mathscr{C}(\mathbb{H})$:

(1) *The equation*

$$\mathrm{d}\eta(t) = \Big\{A\eta(t) + \big((\nabla_{\eta_t} b(t,\cdot)(X_t^\xi) - \frac{\eta(t)}{u(t)}\big)1_{[0,T-r_0)}(t)\Big\}\mathrm{d}t \tag{4.48}$$
$$+((\nabla_{\eta(t)}\sigma)(t,X^\xi(t)))\mathrm{d}W(t), \ \ \eta_0 = \eta,$$

has a unique solution such that $\eta(t) = 0$ *for* $t \geq T - r_0$.

(2) *If* $\|\sigma^{-1}(t,\cdot)\| \leq c(1+|\cdot|^q)$ *holds for some constants* $c, q > 0$ *and all* $t \in [0,T]$, *then*

$$(\nabla_\eta P_T f)(\xi) = \mathbb{E}\bigg(f(X_T^\xi)\int_0^T \Big\langle \sigma^{-1}(t,X^\xi(t))\Big\{\frac{\eta(t)}{u(t)}1_{[0,T-r_0)}(t) \tag{4.49}$$
$$+(\nabla_{\eta_t} b(t,\cdot)(X_t^\xi)1_{[T-r_0,T]}(t)\Big\}, \mathrm{d}W(t)\Big\rangle\bigg)$$

holds for $f \in C_b^1(\mathscr{C}(\mathbb{H}))$.

Proof. (1) By **(A4.5)**, it is easy to see that (4.48) has a unique solution for $t \in [0, T-r_0)$. Let

$$\tilde\eta(t) = \eta(t)1_{[-r_0,T-r_0)}(t), \quad t \geq -r_0.$$

If

$$\lim_{t\uparrow T-r_0} \eta(t) = 0, \tag{4.50}$$

then it is easy to see that $(\tilde\eta(t))_{t\geq 0}$ solves (4.48), and hence the proof is finished. By Itô's formula and (4.52) below, we can deduce that

$$\begin{aligned}\mathrm{d}\frac{|\eta(t)|^p}{u^{p-1}(t)} &= \frac{1}{u^{p-1}(t)}\mathrm{d}|\eta(t)|^p - (p-1)\frac{u'(t)\|\eta(t)\|_H^p}{u^p(t)}\mathrm{d}t\\ &\leq -\theta_p\frac{|\eta(t)|^p}{u^p(t)}\mathrm{d}t + C_1\|\eta(t)\|_\infty^p\mathrm{d}t\\ &\quad+\frac{p}{u^{p-1}(t)}|\eta(t)|^{p-2}\langle\eta(t),((\nabla_{\eta(t)}\sigma(t,\cdot)(X^\xi(t)))\mathrm{d}W(t)\rangle\end{aligned}$$

for some constant $C_1 > 0$. Combining this with Lemma 4.3.8, we obtain

$$\mathbb{E}\int_0^{T-r_0}\frac{|\eta(t)|^p}{u^p(t)}\mathrm{d}t \leq C_2\Big(\|\eta\|_\infty^p + \frac{|\eta(0)|^p}{u^{p-1}(0)}\Big) \tag{4.51}$$

for some constant $C_2 > 0$, and due to the Burkholder–Davis–Gundy inequality,

$$\mathbb{E}\sup_{s\in[0,T-r_0)}\frac{|\eta(s)|^p}{u^{p-1}(s)} < \infty.$$

Since $u(s)\downarrow 0$ as $s \uparrow T - r_0$, the latter implies (4.50).

(2) Let

$$h(t)=\int_0^t \sigma^{-1}(s,X^\xi(s))\Big\{\frac{\eta(s)}{u(s)}1_{[0,T-r_0)}(s)+(\nabla_{\eta_s}b(t,\cdot)(X_s^\xi)1_{[T-r_0,T]}(s)\Big\}\mathrm{d}s,\ t\geq 0.$$

According to the boundedness of $\|\nabla b\|$ and using Hölder's inequality, we obtain

$$\begin{aligned}\mathbb{E}\int_0^T|h'(t)|^2\mathrm{d}t\leq\ &\mathbb{E}\int_0^{T-r_0}\|\sigma^{-1}(t,X^\xi(t))\|^2\frac{|\eta(t)|^2}{u^2(t)}\mathrm{d}t\\ &+C\mathbb{E}\int_{T-r_0}^T\|\sigma^{-1}(t,X^\xi(t))\|^2\|\eta_t\|_\infty^2\mathrm{d}t\\ \leq&\Big(\mathbb{E}\int_0^T\|\sigma^{-1}(t,X^\xi(t))\|^{\frac{2p}{p-2}}\mathrm{d}t\Big)^{\frac{p-2}{p}}\\ &\times\Big\{\Big(\mathbb{E}\int_0^{T-r_0}\frac{|\eta(t)|^p}{u^p(t)}\mathrm{d}t\Big)^{\frac{2}{p}}+C\Big(\mathbb{E}\int_{T-r_0}^T\|\eta_t\|_\infty^p\mathrm{d}t\Big)^{\frac{2}{p}}\Big\}\end{aligned}$$

for some constant $C>0$. Combining this with (4.51), (4.47), $\|\sigma^{-1}(t,x)\|\leq c(1+|x|^q)$, and Lemma 4.3.8 below, we conclude that $\mathbb{E}\int_0^T|h'(t)|^2\mathrm{d}t<\infty$.

Next, we intend to show that $\nabla_\eta X_T^\xi=D_hX_T^\xi$, which implies the desired derivative formula as explained at the beginning of this section. It is easy to see from the definition of h that $\Gamma(t):=\nabla_\eta X^\xi(t)-D_hX^\xi(t)$ solves the equation

$$\begin{aligned}\mathrm{d}\Gamma(t)=&\Big\{A\Gamma(t)+\nabla_{\Gamma_t}b(t,\cdot)(X_t^\xi)-\frac{\eta(t)}{u(t)}1_{[0,T-r_0)}(t)\\ &-(\nabla_{\eta_t}b(t,\cdot)(X_t^\xi)1_{[T-r_0,T]}(t)\Big\}\mathrm{d}t+((\nabla_{\Gamma(t)}\sigma)(t,X^\xi(t)))\mathrm{d}W(t),\ \ t\in[0,T],\Gamma_0=\eta.\end{aligned}$$

Then for $t\in[0,T]$,

$$\begin{aligned}\mathrm{d}(\Gamma(t)-\eta(t))=&\ \Big\{A(\Gamma(t)-\eta(t))+(\nabla_{\Gamma_t-\eta_t}b(t,\cdot)(X_t^\xi)\Big\}\mathrm{d}t\\ &+((\nabla_{\Gamma(t)-\eta(t)}\sigma)(t,X^\xi(t)))\mathrm{d}W(t),\ \ \Gamma_0-\eta_0=0.\end{aligned}$$

By Itô's formula and using **(A4.5)**, we obtain

$$\mathrm{d}|\Gamma(t)-\eta(t)|^2\leq C\|\Gamma_t-\eta_t\|_\infty^2\mathrm{d}t+2\langle\Gamma(t)-\eta(t),((\nabla_{\Gamma(t)-\eta(t)}\sigma)(t,X^\xi(t)))\mathrm{d}W(t)\rangle$$

for some constant $C>0$ and all $t\in[0,T]$. By the boundedness of $\|\nabla\sigma\|_{HS}$ and applying the Burkholder–Davis–Gundy inequality, we obtain

$$\mathbb{E}\sup_{s\in[0,t]}|\Gamma(s)-\eta(s)|^2\leq C'\int_0^t\mathbb{E}\sup_{s\in[0,r]}|\Gamma(s)-\eta(s)|^2\mathrm{d}r,\ \ t\in[0,T],$$

for some constant $C'>0$. Therefore $\Gamma(t)=\eta(t)$ for all $t\in[0,T]$. In particular, $\Gamma_T=\eta_T$. Since $\eta_T=0$, we obtain $\nabla_\eta X_T^\xi=D_hX_T^\xi$. □

Lemma 4.3.8 *In the situation of Theorem* 4.3.7, *let* $(\eta(t))_{t\in[0,T-r_0)}$ *solve* (4.48). *Then for every* $p>0$, *there exists a constant* $C>0$ *such that*

$$\mathbb{E}\sup_{t\in[0,T-r_0)}\|\eta_t\|_\infty^p < C\|\eta\|_\infty^p,\ \ \eta\in\mathscr{C}.$$

Proof. By Itô's formula and the boundedness of ∇b and $\nabla\sigma$, there exists a constant $c_1>0$ such that

$$\begin{aligned}
\mathrm{d}|\eta(t)|^2 &= \Big\{2\Big\langle \eta(t), A\eta(t)+\nabla_{\eta_t}b(t,\cdot)(X_t^\xi)-\frac{\eta(t)}{u(t)}\Big\rangle+\|(\nabla_{\eta(t)}\sigma)(X^\xi(t))\|_{HS}^2\Big\}\mathrm{d}t\\
&\quad +2\langle\eta(t),(\nabla_{\eta(t)}\sigma(t,\cdot)(X^\xi(t))\mathrm{d}W(t)\rangle\\
&\le \Big\{c_1\|\eta_t\|_\infty^2-\frac{2|\eta(t)|^2}{u(t)}1_{[0,T-r_0)}(t)\Big\}\mathrm{d}t+2\langle\eta(t),((\nabla_{\eta(t)}\sigma(t,\cdot)(X^\xi(t)))\mathrm{d}W(t)\rangle
\end{aligned}$$

holds for $t\in[0,T-r_0)$. So for $p>0$, there exists a constant $c_2>0$ such that

$$\begin{aligned}
\mathrm{d}|\eta(t)|^p &= \mathrm{d}(|\eta(t)|^2)^{\frac{p}{2}}=\frac{p}{2}|\eta(t)|^{p-2}\mathrm{d}|\eta(t)|^2\\
&\quad +\frac{p}{2}(p-2)|\eta(t)|^{p-4}\big|((\nabla_{\eta(t)}\sigma(t,\cdot)(X^\xi(t)))^*\eta(t)\big|^2\mathrm{d}t \qquad (4.52)\\
&\le \Big\{c_2\|\eta_t\|_\infty^p-\frac{p|\eta(t)|^p}{u(t)}\Big\}\mathrm{d}t+p|\eta(t)|^{p-2}\langle\eta(t),((\nabla_{\eta(t)}\sigma(t,\cdot)(X^\xi(t)))\mathrm{d}W(t)\rangle
\end{aligned}$$

holds for $t\in[0,T-r_0)$. Since $\|\nabla_{\eta(t)}\sigma(X^\xi(t))\|_{HS}\le c|\eta(t)|$ holds for some constant $c>0$, combining this with the Burkholder–Davis–Gundy inequality, we arrive at

$$\mathbb{E}\sup_{s\in[-r_0,t]}|\eta(s)|^p\le\|\eta\|_\infty^p+c_3\int_0^t\mathbb{E}\sup_{s\in[-r_0,\theta]}|\eta(s)|^p\mathrm{d}\theta,\ \ t\in[0,T-r_0),$$

for some constant $c_3>0$. The proof is then completed by Gronwall's lemma. □

4.4 Stochastic Functional Hamiltonian System

As a typical model of degenerate diffusion systems, the stochastic Hamiltonian system has been investigated recently in [19, 70, 75] using coupling and the Malliavin calculus to establish the Harnack inequalities and the derivative formula. In this section we aim to investigate the functional version of this model (see [7]).

Let $m\in\mathbb{Z}_+$ and $d\in\mathbb{N}$. Set $\mathbb{R}^{m+d}=\mathbb{R}^m\times\mathbb{R}^d$, where $\mathbb{R}^m=\{0\}$ when $m=0$. For $r_0>0$, let $\mathscr{C}=\mathscr{C}(\mathbb{R}^{m+d}):=C([-r_0,0];\mathbb{R}^{m+d})$. Consider the following functional SDE on $\mathbb{R}^{m+d}$:

$$\begin{cases}\mathrm{d}X(t)=\{AX(t)+MY(t)\}\mathrm{d}t,\\ \mathrm{d}Y(t)=\{Z(X(t),Y(t))+b(X_t,Y_t)\}\mathrm{d}t+\sigma\mathrm{d}B(t),\end{cases}\qquad(4.53)$$

where $B(t)$ is a d-dimensional Brownian motion, σ is an invertible $d\times d$ matrix, A is an $m\times m$ matrix, M is an $m\times d$ matrix, $Z:\mathbb{R}^m\times\mathbb{R}^d\to\mathbb{R}^d$ and $b:\mathscr{C}\to\mathbb{R}^d$ are locally Lipschitz continuous (i.e., Lipschitz on compact sets), $(X_t,Y_t)_{t\geq 0}$ is a process on $\mathscr{C}$ with $(X_t,Y_t)(\theta):=(X(t+\theta),Y(t+\theta)),\theta\in[-r_0,0]$. To ensure that P_tf is differentiable for every bounded measurable function f and $t>0$, we will need a rank assumption on A and M such that the noise part of Y_t can also smooth the distribution of X_t via the linear drift terms. More precisely, we will make use of the following Hörmander-type rank condition: there exists an integer $0\leq k\leq m-1$ such that

$$\text{Rank}[M,AM,\ldots,A^kM]=m. \tag{4.54}$$

When $m=0$, this condition automatically holds by convention. Note that when $m\geq 1$, this rank condition holds for some $k>m-1$ if and only if it holds for $k=m-1$.

Let

$$\begin{aligned}Lf(x,y):=\ &\langle Ax+My,\nabla^{(1)}f(x,y)\rangle+\langle Z(x,y),\nabla^{(2)}f(x,y)\rangle\\&+\frac{1}{2}\sum_{i,j=1}^d(\sigma\sigma^*)_{ij}\frac{\partial^2}{\partial y_i\partial y_j}f(x,y),\ \ (x,y)\in\mathbb{R}^{m+d},f\in C^2(\mathbb{R}^{m+d}).\end{aligned}$$

We assume the following:

(A4.6) There exist constants $\lambda,l>0$ and $W\in C^2(\mathbb{R}^{m+d})$ of compact level sets with $W\geq 1$ such that

$$LW\leq\lambda W,\ \ |\nabla^{(2)}W|\leq\lambda W; \tag{4.55}$$
$$\langle b(\xi),\nabla^{(2)}W(\xi(0))\rangle\leq\lambda\|W(\xi)\|_\infty,\ \ \xi\in\mathscr{C}; \tag{4.56}$$
$$|Z(z)-Z(z')|\leq\lambda|z-z'|W(z')^l,\ \ z,z'\in\mathbb{R}^{m+d},|z-z'|\leq 1; \tag{4.57}$$
$$|b(\xi)-b(\xi')|\leq\lambda\|\xi-\xi'\|_\infty\|W(\xi')\|_\infty^l,\xi,\xi'\in\mathscr{C},\|\xi-\xi'\|_\infty\leq 1. \tag{4.58}$$

It is easy to see that **(A4.6)** holds for $W(z)=1+|z|^2$, $l=1$, and some constant $\lambda>0$, provided that Z and b are globally Lipschitz continuous on $\mathbb{R}^{m+d}$ and $\mathscr{C}$ respectively. It is clear that (4.55) and (4.56) imply the nonexplosion of the solution (see Lemma 4.4.6 below). Let P_t be the semigroup of the segment solution, i.e.,

$$P_tf(\xi)=\mathbb{E}^\xi f(X_t,Y_t),\ \ f\in\mathscr{B}_b(\mathscr{C}),\ \xi\in\mathscr{C},$$

where $\mathbb{E}^\xi$ stands for the expectation for the solution starting at the point $\xi\in\mathscr{C}$.

4.4.1 Main Result and Consequences

The following result provides an explicit derivative formula for P_T, $T>r_0$.

Theorem 4.4.1 *Assume* **(A4.6)** *and let* $T > r_0$ *and* $h = (h_1, h_2) \in \mathscr{C}$ *be fixed. Let* $v : [0,T] \to \mathbb{R}$ *and* $\alpha : [0,T] \to \mathbb{R}^m$ *be Lipschitz continuous such that* $v(0) = 1$, $\alpha(0) = 0$, $v(s) = 0$, $\alpha(s) = 0$, *for* $s \geq T - r_0$, *and*

$$h_1(0) + \int_0^t e^{-sA} M\phi(s)ds = 0, \quad t \geq T - r_0, \tag{4.59}$$

where $\phi(s) := v(s)h_2(0) + \alpha(s)$. *Then for* $f \in \mathscr{B}_b(\mathscr{C})$,

$$\nabla_h P_T f(\xi) = \mathbb{E}^\xi \left\{ f(X_T, Y_T) \int_0^T \langle \sigma^{-1} N(s), dB(s) \rangle \right\}, \quad \xi \in \mathscr{C}, \tag{4.60}$$

holds for

$$N(s) := (\nabla_{h(s)} Z)(X(s), Y(s)) + (\nabla_{h_s} b)(X_s, Y_s) - v'(s)h_2(0) - \alpha'(s), \quad s \in [0,T],$$

where h is extended to $[-r_0, T]$ *by letting*

$$h(s) = \left(e^{As} h_1(0) + \int_0^s e^{(s-r)A} M\phi(r)dr, \ \phi(s)\right), \quad s \in (0,T].$$

To apply Theorem 4.4.1, a simple choice of v is

$$v(s) = \frac{(T - r_0 - s)^+}{T - r_0}, \quad s \geq 0.$$

To present a specific choice of α, let

$$Q_t := \int_0^t \frac{s(T - r_0 - s)^+}{(T - r_0)^2} e^{-sA} MM^* e^{-sA^*} ds, \quad t > 0.$$

According to [45] (see also [70, Proof of Theorem 4.2(1)]), when $m \geq 1$, the condition (4.54) implies that Q_t is invertible with

$$\|Q_t^{-1}\| \leq c(T - r_0)(t \wedge 1)^{-2(k+1)}, \quad t > 0, \tag{4.61}$$

for some constant $c > 0$.

Corollary 4.4.2 *Assume* **(A4.6)** *and let* $T > r_0$. *If* (4.54) *holds for some* $0 \leq k \leq m - 1$, *then* (4.60) *holds for* $v(s) = \frac{(T-r_0-s)^+}{T-r_0}$ *and*

$$\alpha(s) = -\frac{s(T - r_0 - s)^+}{(T - r_0)^2} M^* e^{-sA^*} Q_{T-r_0}^{-1} \left(h_1(0) + \int_0^{T-r_0} \frac{T - r_0 - r}{T - r_0} e^{-rA} M h_2(0) dr \right),$$

where by convention, $M = 0$ *(whence* $\alpha = 0$*) if* $m = 0$.

The following gradient estimates are direct consequences of Theorem 4.4.1.

Corollary 4.4.3 *Assume* **(A4.6)**. *If* (4.54) *holds for some* $0 \leq k \leq m - 1$, *then:*

(1) *There exists a constant $C \in (0,\infty)$ such that*

$$|\nabla_h P_T f(\xi)| \le C \sqrt{P_T f^2(\xi)} \Big\{ |h(0)| \Big(\frac{1}{\sqrt{T-r_0}} + \frac{\|M\|}{(T-r_0)^{2k+3/2} \wedge 1} \Big) + \|W(\xi)\|_\infty^l \sqrt{T \wedge (1+r_0)} \Big(\|h\|_\infty + \frac{\|M\| \cdot |h(0)|}{(T-r_0)^{2k+1} \wedge 1} \Big) \Big\}$$

holds for all $T > r_0$, $\xi, h \in \mathscr{C}$, and $f \in \mathscr{B}_b(\mathscr{C})$.

(2) *Let $|\nabla^{(2)} W|^2 \le \delta W$ hold for some constant $\delta > 0$. If $l \in [0, \frac{1}{2})$, then there exists a constant $C \in (0,\infty)$ such that*

$$\begin{aligned} |\nabla_h P_T f(\xi)| &\le r\{P_T f \log f - (P_T f)\log P_T f\}(\xi) \\ &+ \frac{C P_T f(\xi)}{r} \Big\{ |h(0)|^2 \Big(\frac{1}{(T-r_0)\wedge 1} + \frac{\|M\|^2}{\{(T-r_0)\wedge 1\}^{4k+3}} \Big) \\ &+ \|h\|_\infty^2 \|W(\xi)\|_\infty + \Big(\|h\|_\infty^2 + \frac{|h(0)|^2 \|M\|^2}{\{(T-r_0)\wedge 1\}^{4k+2}} \Big)^{\frac{1}{1-2l}} \Big(\frac{r^2}{\|h\|_\infty^2} \wedge r \Big)^{-\frac{4l}{1-2l}} \Big\} \end{aligned}$$

holds for all $r > 0$, $T > r_0$, $\xi, h \in \mathscr{C}$, and positive $f \in \mathscr{B}_b(\mathscr{C})$.

(3) *Let $|\nabla^{(2)} W|^2 \le \delta W$ hold for some constant $\delta > 0$. If $l = \frac{1}{2}$, then there exist constants $C, C' \in (0,\infty)$ such that*

$$\begin{aligned} |\nabla_h P_T f(\xi)| \le\ & r\{P_T f \log f - (P_T f)\log P_T f\}(\xi) \\ &+ \frac{C P_T f(\xi)}{r} \Big\{ |h(0)|^2 \Big(\frac{1}{(T-r_0)\wedge 1} + \frac{\|M\|^2}{\{(T-r_0)\wedge 1\}^{4k+3}} \Big) \\ &+ \|W(\xi)\|_\infty \Big(\|h\|_\infty^2 + \frac{\|M\|^2 |h(0)|^2}{\{(T-r_0)\wedge 1\}^{4k+2}} \Big) \Big\} \end{aligned}$$

holds for all $T > r_0$, $\xi, h \in \mathscr{C}$, positive $f \in \mathscr{B}_b(\mathscr{C})$, and

$$r \ge C' \Big(\|h\|_\infty + \frac{\|M\| \cdot |h(0)|}{\{(T-r_0)\wedge 1\}^{2k+1}} \Big).$$

When $m = 0$, the above assertions hold with $\|M\| = 0$.

According to Proposition 1.3.1 and Corollary 4.4.3(2), we have the following result. Similarly, Corollary 4.4.3(3) implies the same type of Harnack inequality for smaller $\|h\|_\infty$ compared to $T - r_0$.

Corollary 4.4.4 *Assume* **(A4.6)** *and that* (4.54) *holds for some $0 \le k \le m-1$. Let $|\nabla^{(2)} W|^2 \le \delta W$ hold for some constant $\delta > 0$. If $l \in [0, \frac{1}{2})$, then there exists a constant $C \in (0,\infty)$ such that*

$$(P_T f)^p(\xi + h) \le P_T f^p(\xi) \exp\Big[\frac{Cp\|h\|_\infty^2}{p-1} \Big\{ \int_0^1 \|W(\xi + sh)\|_\infty \mathrm{d}s + \frac{1}{(T-r_0)\wedge 1}$$

$$+\frac{\|M\|^2}{(T-r_0)^{4k+3}\wedge 1}+\Big(1\vee\frac{p\|h\|_\infty}{p-1}\Big)^{\frac{4l}{1-2l}}\Big(1+\frac{\|M\|^2}{(T-r_0)^{4k+2}\wedge 1}\Big)^{\frac{1}{1-2l}}\Big\}\Big]$$

holds for all $T>r_0$, $p>1$, $\xi,h\in\mathscr{C}$, and positive $f\in\mathscr{B}_b(\mathscr{C})$. If $m=0$, then the assertion holds for $\|M\|=0$.

Proof. By Corollary 4.4.3(2), we have

$$\sup_{\|h\|_\infty\le 1}|\nabla_h P_T f|(\xi)\le\delta\{P_T(f\log f)-(P_T f)\log P_T f\}(\xi)+\beta(\delta,\xi)P_T f(\xi)$$

for all $\xi\in\mathscr{C}$ and

$$\beta(\delta,\xi)=\frac{C}{\delta}\Big\{\frac{1}{(T-r_0)\wedge 1}+\frac{\|M\|^2}{(T-r_0)^{4k+3}\wedge 1}+\|W(\xi)\|_\infty$$
$$+\Big(1\vee\frac{1}{\delta}\Big)^{\frac{4l}{1-2l}}\Big(1+\frac{\|M\|^2}{(T-r_0)^{4k+2}\wedge 1}\Big)^{\frac{1}{1-2l}}\Big\}$$

for some constant $C>0$. Then the proof is finished by Proposition 1.3.1.

Finally, as a complement to Corollary 4.4.4, where $l\in[0,\frac{1}{2})$ is assumed, we consider the log-Harnack inequality for $l\ge\frac{1}{2}$. To this end, we slightly strengthen (4.57) and (4.58) as follows: there exists an increasing function U on $[0,\infty)$ such that

$$|Z(z)-Z(z')|\le\lambda|z-z'|\{W(z')^l+U(|z-z'|)\};\tag{4.62}$$
$$|b(\xi)-b(\xi')|\le\lambda\|\xi-\xi'\|_\infty\{\|W(\xi')\|_\infty^l+U(\|\xi-\xi'\|_\infty)\}\tag{4.63}$$

for all $z,z'\in\mathbb{R}^{m+d}$ and $\xi,\xi'\in\mathscr{C}$. Obviously, if

$$W(z)^l\le c\{W(z')^l+U(|z-z'|)\},\quad z,z'\in\mathbb{R}^{m+d},$$

holds for some constant $c>0$, then (4.57) and (4.58) imply (4.62) and (4.63) respectively with possibly different λ.

Theorem 4.4.5 *Assume* (4.55), (4.56), (4.62), *and* (4.63). *If* (4.54) *holds for some $0\le k\le m-1$, then there exists a constant $C\in(0,\infty)$ such that for every positive $f\in\mathscr{B}_b(\mathscr{C})$, $T>r_0$, and $\xi,h\in\mathscr{C}$,*

$$P_T\log f(\xi+h)-\log P_T f(\xi)$$
$$\le C\Big\{\Big[\|W(\xi+h)\|_\infty^{2l}+U^2\Big(C\|h\|_\infty+\frac{C\|M\|\cdot|h(0)|}{(T-r_0)\wedge 1}\Big)\Big]\Big(\|h\|_\infty+\frac{\|M\|\cdot|h(0)|}{(T-r_0)^{2k+1}\wedge 1}\Big)^2$$
$$+\frac{|h(0)|^2}{(T-r_0)\wedge 1}+\frac{\|M\|^2|h(0)|^2}{\{(T-r_0)\wedge 1\}^{4k+3}}\Big\}.$$

If $m=0$, then the assertion holds for $\|M\|=0$.

4.4.2 Proof of Theorem 4.4.1

Lemma 4.4.6 *Assume* (4.55) *and* (4.56). *Then for every $k>0$, there exists a constant $C>0$ such that*

$$\mathbb{E}^{\xi}\sup_{-r_0\le s\le t} W(X(s),Y(s))^k \le 3\|W(\xi)\|_\infty^k e^{Ct},\ \ t\ge 0,\ \xi\in\mathscr{C}$$

holds. Consequently, the solution is nonexplosive.

Proof. For $n\ge 1$, let

$$\tau_n=\inf\{t\in[0,T]:|X(t)|+|Y(t)|\ge n\}.$$

Moreover, let

$$\ell(s)=W(X,Y)(s),\ \ s\ge -r_0.$$

By Itô's formula and using (4.55) and (4.56), we may find a constant $C_1>0$ such that

$$\begin{aligned}\ell(t\wedge\tau_n)^k &= \ell(0)^k+k\int_0^{t\wedge\tau_n}\ell(s)^{k-1}\langle\nabla^{(2)}W(X,Y)(s),\sigma\mathrm{d}B(s)\rangle\\ &\quad +k\int_0^{t\wedge\tau_n}\ell(s)^{k-1}\Big\{LW(X,Y)(s)+\big\langle b(X_s,Y_s),\nabla^{(2)}W(X,Y)(s)\big\rangle\\ &\quad +\frac{1}{2}(k-1)\ell(s)^{-1}|\sigma^*\nabla^{(2)}W(X,Y)(s)|^2\Big\}\mathrm{d}s\\ &\le \ell(0)^k+k\int_0^{t\wedge\tau_n}\ell(s)^{k-1}\langle\nabla^{(2)}W(X,Y)(s),\sigma\mathrm{d}B(s)\rangle+C_1\int_0^{t\wedge\tau_n}\sup_{r\in[-r_0,s]}\ell(r)^k\mathrm{d}s.\end{aligned}\tag{4.64}$$

By the second inequality in (4.55) and the Burkholder–Davis–Gundy inequality, we obtain

$$\begin{aligned}&k\mathbb{E}^{\xi}\sup_{s\in[0,t]}\Big|\int_0^{s\wedge\tau_n}\ell(r)^{k-1}\langle\nabla^{(2)}W(X,Y)(s),\sigma\mathrm{d}B(r)\rangle\Big|\le C_2\mathbb{E}^{\xi}\Big(\int_0^t\ell(s\wedge\tau_n)^{2k}\mathrm{d}s\Big)^{\frac12}\\ &\le C_2\mathbb{E}^{\xi}\Big\{\Big(\sup_{s\in[0,t]}\ell(s\wedge\tau_n)^k\Big)^{\frac12}\Big(\int_0^t\ell(s\wedge\tau_n)^k\mathrm{d}s\Big)^{\frac12}\Big\}\\ &\le\frac12\mathbb{E}^{\xi}\sup_{s\in[0,t]}\ell(s\wedge\tau_n)^k+\frac{C_2^2}{2}\mathbb{E}^{\xi}\int_0^t\sup_{r\in[0,s]}\ell(r\wedge\tau_n)^k\mathrm{d}s\end{aligned}$$

for some constant $C_2>0$. Combining this with (4.64) and noting that $(X_0,Y_0)=\xi$, we conclude that there exists a constant $C>0$ such that

$$\mathbb{E}^{\xi}\sup_{-r_0\le s\le t}\ell(s\wedge\tau_n)^k\le 3\|W(\xi)\|_\infty^k+C\mathbb{E}^{\xi}\int_0^t\sup_{s\in[-r_0,t]}\ell(s)^k\mathrm{d}s,\ \ t\ge 0.$$

Due to Gronwall's lemma, this implies that

$$\mathbb{E}^{\xi} \sup_{-r_0 \le s \le t} \ell(s \wedge \tau_n)^k \le 3\|W(\xi)\|_\infty^k \mathrm{e}^{Ct}, \quad t \ge 0, n \ge 1.$$

Consequently, we have $\tau_n \uparrow \infty$ as $n \uparrow \infty$, and thus the desired inequality follows by letting $n \to \infty$. □

To establish the derivative formula, we first construct couplings for solutions starting from ξ and $\xi + \varepsilon h$ for $\varepsilon \in (0,1]$, then let $\varepsilon \to 0$. For fixed $\xi = (\xi_1, \xi_2)$, $h = (h_1, h_2) \in \mathscr{C}$, let $(X(t), Y(t))$ solve (4.53) with $(X_0, Y_0) = \xi$; and for $\varepsilon \in (0,1]$, let $(X^\varepsilon(t), Y^\varepsilon(t))$ solve the equation

$$\begin{aligned}
&\mathrm{d}X^\varepsilon(t) = \{AX^\varepsilon(t) + MY^\varepsilon(t)\}\mathrm{d}t, \\
&\mathrm{d}Y^\varepsilon(t) = \{Z(X(t), Y(t)) + b(X_t, Y_t)\}\mathrm{d}t + \sigma \mathrm{d}B(t) + \varepsilon\{v'(t)h_2(0) + \alpha'(t)\}\mathrm{d}t
\end{aligned} \tag{4.65}$$

with $(X_0^\varepsilon, Y_0^\varepsilon) = \xi + \varepsilon h$. By Lemma 4.4.6 and (4.66) below, the solution to (4.65) is nonexplosive as well.

Proposition 4.4.7 *Let $\phi(s) = v(s)h_2(0) + \alpha(s)$, $s \in [0,T]$, and the conditions of Theorem 4.4.1 hold. Then*

$$(X^\varepsilon(t), Y^\varepsilon(t)) = (X(t), Y(t)) + \varepsilon h(t), \quad \varepsilon, t \ge 0, \tag{4.66}$$

holds for

$$h(t) = \left(\mathrm{e}^{At}h_1(0) + \int_0^t \mathrm{e}^{(t-r)A} M\phi(r)\mathrm{d}r,\ \phi(t)\right), \quad t \in [0,T].$$

In particular, $(X_T^\varepsilon, Y_T^\varepsilon) = (X_T, Y_T)$.

Proof. By (4.65) and noting that $v(0) = 1$ and $v(s) = 0$ for $s \ge T - r_0$, we have $Y^\varepsilon(t) = Y(t) + \varepsilon\phi(t)$ and

$$X^\varepsilon(t) = X(t) + \varepsilon \mathrm{e}^{At}h_1(0) + \varepsilon \int_0^t \mathrm{e}^{(t-s)A} M\phi(s)\mathrm{d}s, \quad t \ge 0.$$

Thus, (4.66) holds. Moreover, since $\alpha(s) = v(s) = 0$ for $s \ge T - r_0$, we have $h^{(2)}(s) = \phi(s) = 0$ for $s \ge T - r_0$. Moreover, by (4.59), we have $h^{(1)}(s) = 0$ for $s \ge T - r_0$. Therefore, the proof is finished. □

Let

$$\begin{aligned}
\Phi^\varepsilon(s) &= Z(X(s), Y(s)) - Z(X^\varepsilon(s), Y^\varepsilon(s)) \\
&\quad + b(X_s, Y_s) - b(X_s^\varepsilon, Y_s^\varepsilon) + \varepsilon\{v'(s)h_2(0) + \alpha'(s)\}, \\
R^\varepsilon(s) &= \exp\left[-\int_0^s \langle \sigma^{-1}\Phi^\varepsilon(r), \mathrm{d}B(r)\rangle - \frac{1}{2}\int_0^s |\sigma^{-1}\Phi^\varepsilon(r)|^2 \mathrm{d}r\right], \\
B^\varepsilon(t) &= B(t) + \int_0^t \sigma^{-1}\Phi^\varepsilon(s)\mathrm{d}s.
\end{aligned}$$

According to Girsanov's theorem, to ensure that $B^\varepsilon(t)$ is a Brownian motion under $\mathbb{Q}_\varepsilon := R^\varepsilon(T)\mathbb{P}$, we first prove that $R^\varepsilon(t)$ is an exponential martingale. To derive the derivative formula using Theorem 1.1.2, we need to prove the uniform integrability of $\{\frac{R^\varepsilon(T)-1}{\varepsilon}\}_{\varepsilon\in(0,1)}$, which is ensured by the following two lemmas.

Lemma 4.4.8 *Let* **(A4.6)** *hold. Then there exists* $\varepsilon_0 > 0$ *such that*

$$\sup_{s\in[0,T],\varepsilon\in(0,\varepsilon_0)} \mathbb{E}[R^\varepsilon(s)\log R^\varepsilon(s)] < \infty,$$

so that for each $\varepsilon \in (0,1)$, $(R^\varepsilon(s))_{s\in[0,T]}$ *is a uniformly integrable martingale.*

Proof. By the definition of h, there exists $\varepsilon_0 > 0$ such that

$$\varepsilon_0|h(t)| \le 1, \quad t \in [-r_0, T]. \tag{4.67}$$

Define

$$\tau_n = \inf\{t \ge 0 : |X(t)| + |Y(t)| \ge n\}, \quad n \ge 1.$$

We have $\tau_n \uparrow \infty$ as $n \uparrow \infty$ due to the nonexplosion. By Girsanov's theorem, the process $\{R^\varepsilon(s\wedge\tau_n)\}_{s\in[0,T]}$ is a martingale and $\{B^\varepsilon(s)\}_{s\in[0,T\wedge\tau_n]}$ is a Brownian motion under the probability measure $\mathbb{Q}_{\varepsilon,n} := R^\varepsilon(T\wedge\tau_n)\mathbb{P}$. By the definition of $R^\varepsilon(s)$, we have

$$\begin{aligned}\mathbb{E}[R^\varepsilon(s\wedge\tau_n)\log R^\varepsilon(s\wedge\tau_n)] &= \mathbb{E}_{\mathbb{Q}_{\varepsilon,n}}[\log R^\varepsilon(s\wedge\tau_n)] \\ &\le \frac{1}{2}\mathbb{E}_{\mathbb{Q}_{\varepsilon,n}}\int_0^{T\wedge\tau_n} |\sigma^{-1}\Phi^\varepsilon(r)|^2 \mathrm{d}r.\end{aligned} \tag{4.68}$$

By (4.67), (4.57), and (4.58),

$$|\sigma^{-1}\Phi^\varepsilon(s)|^2 \le c\varepsilon^2\|W(X_s^\varepsilon, Y_s^\varepsilon)\|_\infty^{2l} \tag{4.69}$$

holds for some constant c independent of ε. Since the distribution of $(X^\varepsilon(s), Y^\varepsilon(s))_{s\in[0,T\wedge\tau_n]}$ under $\mathbb{Q}_{\varepsilon,n}$ coincides with that of the solution to (4.53) with $(X_0, Y_0) = \xi + \varepsilon h$ up to time $T\wedge\tau_n$, we therefore obtain from Lemma 4.4.6 that

$$\mathbb{E}[R^\varepsilon(s\wedge\tau_n)\log R^\varepsilon(s\wedge\tau_n)] \le c\|W(\xi+\varepsilon h)\|_\infty^{2l}\int_0^T \mathrm{e}^{Ct}\mathrm{d}t < \infty, \quad n \ge 1, \varepsilon \in (0,\varepsilon_0).$$

Then the required assertion follows by letting $n \to \infty$. □

Lemma 4.4.9 *If* **(A4.6)** *holds, then there exists* $\varepsilon_0 > 0$ *such that*

$$\sup_{\varepsilon\in(0,\varepsilon_0)} \mathbb{E}\left(\frac{R^\varepsilon(T)-1}{\varepsilon}\log\frac{R^\varepsilon(T)-1}{\varepsilon}\right) < \infty.$$

Moreover,

$$\lim_{\varepsilon\to 0}\frac{R^\varepsilon(T)-1}{\varepsilon}= \tag{4.70}$$
$$\int_0^T \left\langle \sigma^{-1}\Big\{(\nabla_{h(s)}Z)(X(s),Y(s))+(\nabla_{h_s}b)(X_s,Y_s)-v'(s)h_2(0)-\alpha'(s)\Big\},\mathrm{d}B(s)\right\rangle.$$

Proof. Let ε_0 be such that (4.67) holds. Since (4.70) is a direct consequence of (4.66) and the definition of $R^\varepsilon(T)$, we prove only the first assertion. It is proved in [19] that

$$\frac{R^\varepsilon(T)-1}{\varepsilon}\log\frac{R^\varepsilon(T)-1}{\varepsilon}\le 2R^\varepsilon(T)\left(\frac{\log R^\varepsilon(T)}{\varepsilon}\right)^2.$$

Since due to Lemma 4.4.8, $\{B^\varepsilon(t)\}_{t\in[0,T]}$ is a Brownian motion under the probability measure $\mathbb{Q}_\varepsilon := R^\varepsilon(T)\mathbb{P}$, and since

$$\begin{aligned}\log R^\varepsilon(T) &= -\int_0^T\langle\sigma^{-1}\Phi^\varepsilon(r),\mathrm{d}B(r)\rangle-\frac{1}{2}\int_0^T|\sigma^{-1}\Phi^\varepsilon(r)|^2\mathrm{d}r\\ &= -\int_0^T\langle\sigma^{-1}\Phi^\varepsilon(r),\mathrm{d}B^\varepsilon(r)\rangle+\frac{1}{2}\int_0^T|\sigma^{-1}\Phi^\varepsilon(r)|^2\mathrm{d}r, \end{aligned}\tag{4.71}$$

it follows from (4.69) that

$$\begin{aligned}&\mathbb{E}\left(\frac{R^\varepsilon(T)-1}{\varepsilon}\log\frac{R^\varepsilon(T)-1}{\varepsilon}\right)\\ &\le\mathbb{E}\left(2R^\varepsilon(T)\left(\frac{\log R^\varepsilon(T)}{\varepsilon}\right)^2\right)=2\mathbb{E}_{\mathbb{Q}_\varepsilon}\left(\frac{\log R^\varepsilon(T)}{\varepsilon}\right)^2\\ &\le\frac{4}{\varepsilon^2}\mathbb{E}_{\mathbb{Q}_\varepsilon}\left(\int_0^T\langle\sigma^{-1}\Phi^\varepsilon(r),\mathrm{d}B^\varepsilon(r)\rangle\right)^2+\frac{1}{\varepsilon^2}\mathbb{E}_{\mathbb{Q}_\varepsilon}\left(\int_0^T|\sigma^{-1}\Phi^\varepsilon(r)|^2\mathrm{d}r\right)^2\\ &\le\frac{4}{\varepsilon^2}\int_0^T\mathbb{E}_{\mathbb{Q}_\varepsilon}|\sigma^{-1}\Phi^\varepsilon(r)|^2\mathrm{d}r+\frac{T}{\varepsilon^2}\int_0^T\mathbb{E}_{\mathbb{Q}_\varepsilon}|\sigma^{-1}\Phi^\varepsilon(r)|^4\mathrm{d}r\\ &\le c\int_0^T\mathbb{E}_{\mathbb{Q}_\varepsilon}\|W(X_r^\varepsilon,Y_r^\varepsilon)\|_\infty^{4l}\mathrm{d}r\end{aligned}$$

holds for some constant $c>0$. As explained in the proof of Lemma 4.4.8, the distribution of $(X_s^\varepsilon,Y_s^\varepsilon)_{s\in[0,T]}$ under $\mathbb{Q}_\varepsilon$ coincides with that of the segment process of the solution to (4.53) with $(X_0,Y_0)=\xi+\varepsilon h$. The first assertion follows by Lemma 4.4.6. □

Proof of Theorem 4.4.1. By Lemma 4.4.9 and the dominated convergence theorem, we have

$$\lim_{\varepsilon\to 0}\frac{R^\varepsilon(T)-1}{\varepsilon}=\int_0^T\langle\sigma^{-1}N(s),\mathrm{d}B(s)\rangle$$

in $L^1(\mathbb{P})$. Then the desired derivative formula follows from Theorem 1.1.2. □

4.4.3 Proofs of Corollary 4.4.3 and Theorem 4.4.5

To prove the entropy–gradient estimates in Corollary 4.4.3 (2) and (3), we need the following simple lemma.

Lemma 4.4.10 *Let $\ell(t)$ be a nonnegative continuous semimartingale and let $\mathscr{M}(t)$ be a continuous martingale with $\mathscr{M}(0) = 0$ such that*

$$\mathrm{d}\ell(t) \le \mathrm{d}\mathscr{M}(t) + c\bar{\ell}_t \mathrm{d}t,$$

where $c \ge 0$ is a constant and $\bar{\ell}_t := \sup_{s\in[0,t]} \ell(s)$. Then

$$\mathbb{E}\exp\left[\frac{c\varepsilon}{\mathrm{e}^{cT}-1}\int_0^T \bar{\ell}_t \mathrm{d}t\right] \le \mathrm{e}^{\varepsilon\ell(0)+1}\left(\mathbb{E}\mathrm{e}^{2\varepsilon^2\langle\mathscr{M}\rangle(T)}\right)^{\frac{1}{2}}, \quad T,\varepsilon \ge 0.$$

Proof. Let $\bar{\mathscr{M}}_t = \sup_{s\in[0,t]} \mathscr{M}(t)$ and $y(t) = \int_0^t \bar{\ell}_s \mathrm{d}s$. We have

$$y'(t) \le \ell(0) + cy(t) + \bar{\mathscr{M}}_t.$$

So

$$y(t) \le \mathrm{e}^{ct}\int_0^t \mathrm{e}^{-cs}\{\ell(0) + \bar{\mathscr{M}}_s\}\mathrm{d}s \le \frac{\mathrm{e}^{ct}-1}{c}\{\ell(0) + \bar{\mathscr{M}}_t\}.$$

Combining this with

$$\mathbb{E}\mathrm{e}^{\varepsilon\bar{\mathscr{M}}_t} \le \mathbb{E}\mathrm{e}^{1+\varepsilon\mathscr{M}(T)} \le \mathrm{e}\left(\mathbb{E}\mathrm{e}^{2\varepsilon^2\langle\mathscr{M}\rangle(T)}\right)^{\frac{1}{2}},$$

we complete the proof. □

Corollary 4.4.11 *Assume* **(A4.6)** *and let $|\nabla^{(2)}W|^2 \le \delta W$ hold for some constant $\delta > 0$. Then there exists a constant $c > 0$ such that*

$$\begin{aligned}
&\mathbb{E}^\xi \exp\left[\frac{c^2}{2\|\sigma\|^2\delta(\mathrm{e}^{cT}-1)^2}\int_0^T \|W(X_t,Y_t)\|_\infty \mathrm{d}t\right]\\
&\le \exp\left[2 + \frac{cW(\xi(0))}{\|\sigma\|^2\delta(\mathrm{e}^{cT}-1)} + \frac{c^2 r_0\|W(\xi)\|_\infty}{2\|\sigma\|^2\delta(\mathrm{e}^{cT}-1)^2}\right], \quad T > r_0.
\end{aligned}$$

Proof. By **(A4.6)** and Itô's formula, there exists a constant $c > 0$ such that

$$\mathrm{d}W(X,Y)(s) \le \langle\nabla^{(2)}W(X,Y)(s), \sigma\mathrm{d}B(s)\rangle + c\|W(X_s,Y_s)\|_\infty \mathrm{d}s.$$

Let

$$\mathscr{M}(t) := \int_0^t \langle\nabla^{(2)}W(X,Y)(s), \sigma\mathrm{d}B(s)\rangle, \quad \ell(t) := W(X,Y)(t),$$

and let $\varepsilon = \frac{c}{2\|s\|^2\delta(\mathrm{e}^{cT}-1)}$ be such that

$$\frac{c\varepsilon}{\mathrm{e}^{cT}-1} = 2\|\sigma\|^2\varepsilon^2\delta.$$

Then by Lemma 4.4.10 and $|\nabla^{(2)}W|^2 \le \delta W$, we have

$$\mathbb{E}^\xi \exp\left[\frac{c\varepsilon}{e^{cT}-1}\int_0^T \bar{\ell}_t \mathrm{d}t\right] \le e^{\varepsilon\ell(0)+1}\left(\mathbb{E}^\xi e^{2\varepsilon^2\langle \mathscr{M}\rangle(T)}\right)^{\frac{1}{2}}$$
$$\le e^{1+\varepsilon\ell(0)}\left(\mathbb{E}^\xi e^{2\varepsilon^2\|\sigma\|^2\delta\int_0^T \bar{\ell}_t \mathrm{d}t}\right)^{\frac{1}{2}} = e^{1+\varepsilon\ell(0)}\left(\mathbb{E}^\xi e^{\frac{c\varepsilon}{e^{cT}-1}\int_0^T \bar{\ell}_t \mathrm{d}t}\right)^{\frac{1}{2}}.$$

Using stopping times as in the proof of Lemma 4.4.6, we may assume that

$$\mathbb{E}^\xi \exp\left[\frac{c\varepsilon}{e^{cT}-1}\int_0^T \bar{\ell}_t \mathrm{d}t\right] < \infty,$$

so that

$$\mathbb{E}^\xi \exp\left[\frac{c\varepsilon}{e^{cT}-1}\int_0^T \bar{\ell}_t \mathrm{d}t\right] \le e^{2+2\varepsilon\ell(0)}.$$

This completes the proof by noting that

$$\frac{c^2}{2\|\sigma\|^2\delta(e^{cT}-1)^2}\int_0^T \|W(X_t,Y_t)\|_\infty \mathrm{d}t \le \frac{c^2 r_0\|W(\xi)\|_\infty}{2\|\sigma\|^2\delta(e^{cT}-1)^2} + \frac{c\varepsilon}{e^{cT}-1}\int_0^T \bar{\ell}_t \mathrm{d}t.$$

□

Proof of Corollary 4.4.3. Let v and α be as given in Corollary 4.4.2. By the semi-group property and Jensen's inequality, we will consider only $T - r_0 \in (0,1]$.

(1) By (4.61) and the definitions of α and v, there exists a constant $C > 0$ such that for every $s \in [0,T]$,

$$|v'(s)h_2(0) + \alpha'(s)| \le C 1_{[0,T-r_0]}(s)|h(0)|\left(\frac{1}{T-r_0} + \frac{\|M\|}{(T-r_0)^{2(k+1)}}\right) \quad (4.72)$$

$$|h(s)| \le C|h(0)|\left(1 + \frac{\|M\|}{(T-r_0)^{2k+1}}\right), \quad (4.73)$$

$$\|h_s\|_\infty \le C\left(\|h\|_\infty + \frac{\|M\|\cdot|h(0)|}{(T-r_0)^{2k+1}}\right). \quad (4.74)$$

Therefore, it follows from (4.57) and (4.58) that

$$|N(s)| \le C 1_{[0,T-r_0]}(s)|h(0)|\left(\frac{1}{T-r_0} + \frac{\|M\|}{(T-r_0)^{2(k+1)}}\right) \quad (4.75)$$
$$+C\left(\|h\|_\infty + \frac{\|M\|\cdot|h(0)|}{(T-r_0)^{2k+1}}\right)\|W(X_s,Y_s)\|_\infty^l$$

holds for some constant $C > 0$. Combining this with Theorem 4.4.1, we obtain

$$|\nabla_h P_T f(\xi)| \le C\sqrt{P_T f^2(\xi)}\left(\mathbb{E}^\xi \int_0^T |N(s)|^2 \mathrm{d}s\right)^{\frac{1}{2}}$$

$$\leq C\sqrt{P_T f^2(\xi)}\bigg\{|h(0)|\Big(\frac{1}{\sqrt{T-r_0}}+\frac{\|M\|}{(T-r_0)^{2k+3/2}}\Big)$$
$$+\Big(\|h\|_\infty+\frac{\|M\|\cdot|h(0)|}{(T-r_0)^{2k+1}}\Big)\Big(\int_0^T \mathbb{E}^\xi\|W(X_s,Y_s)\|_\infty^{2l}\mathrm{d}s\Big)^{\frac{1}{2}}\bigg\}.$$

This completes the proof of (1), since due to Lemma 4.4.6 one has

$$\mathbb{E}^\xi\|W(X_s,Y_s)\|_\infty^{2l}\leq 3\|W(\xi)\|_\infty^{2l}\mathrm{e}^{Cs},\quad s\in[0,T]$$

for some constant $C>0$.

(2) By Theorem 4.4.1 and Young's inequality (cf. [4, Lemma 2.4]), we have

$$\begin{aligned}|\nabla_h P_T f|(\xi)\ &\leq r\{P_T f\log f-(P_T f)\log P_T f\}(\xi)\\ &\quad +rP_T f(\xi)\log\mathbb{E}^\xi\mathrm{e}^{\frac{1}{r}\int_0^T\langle N(s),(\sigma^*)^{-1}\mathrm{d}B(s)\rangle},\quad r>0.\end{aligned}\tag{4.76}$$

Next, it follows from (4.75) that

$$\begin{aligned}&\left(\mathbb{E}^\xi\exp\left[\frac{1}{r}\int_0^T\langle N(s),(\sigma^*)^{-1}\mathrm{d}B(s)\rangle\right]\right)^2\leq\mathbb{E}^\xi\exp\left[\frac{2\|\sigma^{-1}\|^2}{r^2}\int_0^T|N(s)|^2\mathrm{d}s\right]\\ &\leq\exp\left[\frac{C_1|h(0)|^2}{r^2}\Big(\frac{1}{T-r_0}+\frac{\|M\|^2}{(T-r_0)^{4k+3}}\Big)\right]\\ &\quad\times\mathbb{E}^\xi\exp\left[\frac{C_1}{r^2}\Big(\|h\|_\infty^2+\frac{\|M\|^2|h(0)|^2}{(T-r_0)^{4k+2}}\Big)\int_0^T\|W(X_s,Y_s)\|_\infty^{2l}\mathrm{d}s\right]\end{aligned}\tag{4.77}$$

holds for some constant $C_1\in(0,\infty)$. Since $2l\in[0,1)$ and $T\leq 1+r_0$, there exists a constant $C_2\in(0,\infty)$ such that

$$\beta\|W(X_s,Y_s)\|_\infty^{2l}\leq\frac{c^2(\frac{\|h\|_\infty^2}{r^2}\wedge 1)\|W(X_s,Y_s)\|_\infty}{2\|\sigma\|^2\delta(\mathrm{e}^{cT}-1)^2}+C_2\beta^{\frac{1}{1-2l}}\Big(\frac{\|h\|_\infty^2}{r^2}\wedge 1\Big)^{-\frac{2l}{1-2l}},\ \beta>0.$$

Taking

$$\beta=\frac{C_1}{r^2}\Big(\|h\|_\infty^2+\frac{\|M\|^2|h(0)|^2}{(T-r_0)^{4k+2}}\Big)$$

and applying Corollary 4.4.11, we arrive at

$$\begin{aligned}\mathbb{E}^\xi\exp\left[\beta\int_0^T\|W(X_s,Y_s)\|_\infty^{2l}\mathrm{d}s\right]&\leq\exp\left[C_2\beta^{\frac{1}{1-2l}}\Big(\frac{\|h\|_\infty^2}{r^2}\wedge 1\Big)^{-\frac{2l}{1-2l}}\right]\\ &\quad\times\left(\mathbb{E}^\xi\exp\left[\frac{c^2}{2\|\sigma\|^2\delta(\mathrm{e}^{cT}-1)^2}\int_0^T\|W(X_s,Y_s)\|_\infty\mathrm{d}s\right]\right)^{\frac{\|h\|_\infty^2}{r^2}\wedge 1}\\ &\leq\exp\left[\frac{C_3}{r^2}\Big\{\|h\|_\infty^2\|W(\xi)\|_\infty+\Big(\|h\|_\infty^2+\frac{\|M\|^2|h(0)|^2}{(T-r_0)^{4k+2}}\Big)^{\frac{1}{1-2l}}(\|h\|_\infty\wedge r)^{-\frac{4l}{1-2l}}\Big\}\right]\end{aligned}$$

for some constant $C_3 \in (0,\infty)$ and all $T \in (r_0, 1+r_0]$. Therefore, the desired entropy–gradient estimate follows by combining this with (4.76) and (4.77).

(3) Let $C' > 0$ be such that $r \geq C'\left(\|h\|_\infty + \frac{\|M\|\cdot|h(0)|}{(T-r_0)^{2k+1}}\right)$ implies

$$\frac{C_1}{r^2}\left(\|h\|_\infty^2 + \frac{\|M\|^2|h(0)|^2}{(T-r_0)^{4k+2}}\right) \leq \frac{c^2}{2\|\sigma\|^2\delta(e^{cT}-1)^2},$$

so that by Corollary 4.4.11,

$$\begin{aligned}
&\mathbb{E}^\xi \exp\left[\frac{C_1}{r^2}\left(\|h\|_\infty^2 + \frac{\|M\|^2|h(0)|^2}{(T-r_0)^{4k+2}}\right)\int_0^T \|W(X_s,Y_s)\|_\infty^{2l}\mathrm{d}s\right] \\
&\leq \left(\mathbb{E}^\xi e^{\frac{c^2}{2\|\sigma\|^2\delta(e^{cT}-1)^2}\int_0^T \|W(X_s,Y_s)\|_\infty \mathrm{d}s}\right)^{2C_1\|\sigma\|^2\delta\frac{(e^{cT}-1)^2}{c^2}\left(\frac{\|h\|_\infty^2}{r^2}+\frac{\|M\|^2|h(0)|^2}{r^2(T-r_0)^{4k+2}}\right)} \\
&\leq \exp\left[\frac{C\|W(\xi)\|_\infty}{r^2}\left(\|h\|_\infty^2 + \frac{\|M\|^2|h(0)|^2}{(T-r_0)^{4k+2}}\right)\right]
\end{aligned}$$

holds for some constant $C > 0$. The proof is finished by combining this with (4.76) and (4.77). □

Proof of Theorem 4.4.5. Again, we prove the result only for $T \in (r_0, 1+r_0]$. By Lemma 4.4.8 and Girsanov's theorem, $\{B^1(t)\}_{t\in[0,T]}$ is a Brownian motion under the probability $\mathbb{Q}_1 := R^1(T)\mathbb{P}$. Thus,

$$\mathbb{E}(R^1 \log R^1)(T) = \mathbb{E}_{\mathbb{Q}_1} \log R^1(T) = \frac{1}{2}\mathbb{E}_{\mathbb{Q}_1}\int_0^T |\sigma^{-1}\Phi^1(r)|^2 \mathrm{d}r. \tag{4.78}$$

By (4.62), (4.63), (4.66), (4.72)–(4.74), and the definition of Φ^1, we have

$$\begin{aligned}
&|\sigma^{-1}\Phi^1(s)|^2 \leq C_1|h(0)|^2\left(\frac{1}{(T-r_0)^2} + \frac{\|M\|^2}{(T-r_0)^{4(k+1)}}\right)1_{[0,T-r_0]}(s) \\
&+C_1\left\{\|W(X_s^1,Y_s^1)\|_\infty^{2l} + U^2\left(C_1\|h\|_\infty + \frac{C_1\|M\|\cdot|h(0)|}{(T-r_0)^{2k+1}}\right)\right\}\left(\|h\|_\infty + \frac{\|M\|\cdot|h(0)|}{(T-r_0)^{2k+1}}\right)^2
\end{aligned}$$

for some constant $C_1 > 0$. Then the proof is completed by combining this with (4.78), Lemma 4.4.6, and Theorem 1.1.1 (note that $(X^1(s),Y^1(s))$ under $\mathbb{Q}_1$ solves the same equation as (X_s,Y_s) under $\mathbb{P}$). □

Glossary

$\mathscr{B}(E)$ The set of measurable functions on E.

$\mathscr{B}_b(E)$ The set of bounded measurable functions on E.

$\mathscr{B}_b^+(E)$ The set of nonnegative bounded measurable functions on E.

$C(E)$ The set of continuous functions on E.

$C_b(E)$ The set of bounded continuous functions on E.

$C_b^+(E)$ The set of bounded nonnegative continuous functions on E.

$C_0(E)$ The set of continuous functions with compact support on E.

$C^p(E)$ The set of functions on E with continuous derivatives up to order p.

$C_b^p(E)$ The set of functions on E with bounded continuous derivatives up to order p.

$C_0^p(E)$ The set of functions on E with compact support and continuous derivatives up to order p.

$\mathscr{C}(E) = C([-r_0, 0]; E)$ For a metric space E.

D The Malliavin gradient operator with respect to the underlying Brownian motion.

$\mathbb{E}$ The expectation with respect to the underlying probability measure $\mathbb{P}$.

$\mathbb{E}_{\mathbb{Q}}$ The expectation with respect to the (changed) probability measure $\mathbb{Q}$.

$\mathbb{H}$ Hilbert space.

$\mathbb{H}^1$ The Cameron–Martin space over $\mathbb{H}$.

$\mathbb{H}_0^{1,p}(D)$ The Sobolev space on an open domain $D \subset \mathbb{R}^d$, which is the closure of $C_0^\infty(D)$ under the norm $\|f\|_{1,p} := \|f\|_{L^p(\mathbf{m})} + \|\nabla f\|_{L^p(\mathbf{m})}$, where $\mathbf{m}$ is the Lebesgue measure on D.

$\mathscr{L}(\mathbb{H})$ The set of densely defined linear operators on $\mathbb{H}$.

$\mathscr{L}_b(\mathbb{H})$ The set of bounded linear operators on $\mathbb{H}$.

F.-Y. Wang, *Harnack Inequalities for Stochastic Partial Differential Equations*, SpringerBriefs in Mathematics, DOI 10.1007/978-1-4614-7934-5,

$\mathscr{L}_{HS}(\mathbb{H})$ The set of Hilbert–Schmidt operators on $\mathbb{H}$.

$\mathscr{L}_S(\mathbb{H})$ The set of densely defined closed linear operators on $\mathbb{H}$.

X_t The segment process associated to an SDDE with time delay r_0, i.e., $X_t(s) := X(t+s), s \in [-r_0, 0]$.

$|\cdot|$ The norm in the underlying Hilbert space $\mathbb{H}$ or the Euclidean space $\mathbb{R}^d$.

$\|\cdot\|$ The operator norm for linear operators.

$\|\cdot\|_{HS}$ The Hilbert–Schmidt norm of linear operators.

$\|\cdot\|_\sigma$ The intrinsic norm induced by a linear operator σ, i.e., $\|x\|_\sigma = \inf\{|y| : \sigma y = x\}$ and $\inf \emptyset = \infty$ by convention.

$\|\cdot\|_\infty$ The uniform norm, i.e., $\|f\|_\infty := \sup|f|$ for a function f.

$\langle \cdot, \cdot \rangle$ The inner product in the underlying Hilbert space $\mathbb{H}$.

$\langle \cdot, \cdot \rangle_2$ The inner product in $L^2(\mathbf{m})$ for a reference measure $\mathbf{m}$.

${}_{\mathbb{V}^*}\langle \cdot, \cdot \rangle_{\mathbb{V}}$ The dualization between a Banach space $\mathbb{V}$ and its duality $\mathbb{V}^*$ with respect to a Hilbert space into which $\mathbb{V}$ is continuously and densely embedded.

∇ The gradient operator with respect to the underlying space variable.

$a \vee b$ $\max\{a, b\}$.

$a \wedge b$ $\min\{a, b\}$.

PDE Partial differential equation.

SDE Stochastic differential equation.

SDDE Stochastic delay differential equation.

SDPDE Stochastic delay partial differential equation.

SPDE Stochastic partial differential equation.

References

1. S. Aida, H. Kawabi, *Short time asymptotics of a certain infinite dimensional diffusion process,* Stochastic analysis and related topics, VII (Kusadasi, 1998), Progr. Probab., 48, 77–124, Birkhäuser Boston, Boston, MA, 2001.
2. S. Aida, T. Zhang, *On the small time asymptotics of diffusion processes on path groups,* Potential Anal. 16(2002), 67–78.
3. M. Arnaudon, A. Thalmaier, F.-Y. Wang, *Harnack inequality and heat kernel estimates on manifolds with curvature unbounded below,* Bull. Sci. Math. 130(2006), 223–233.
4. M. Arnaudon, A. Thalmaier, F.-Y. Wang, *Gradient estimates and Harnack inequalities on noncompact Riemannian manifolds,* Stochastic Process. Appl. 119(2009), 3653–3670.
5. M. Arnaudon, A. Thalmaier, F.-Y. Wang, *Equivalent Harnack and gradient inequalities for pointwise curvature lower bound,* arXiv:1209.6161.
6. D. Bakry, I. Gentil, M. Ledoux, *On Harnack inequalities and optimal transportation,* arXiv: 1210.4650.
7. J. Bao, F.-Y. Wang, C. Yuan, *Derivative formula and Harnack inequality for degenerate functional SDEs,* to appear in Stochastics and Dynamics. Stoch. Dyn. 13(2013), 1250013, 22 pages.
8. J. Bao, F.-Y. Wang, C. Yuan, *Bismut Formulae and Applications for Functional SPDEs,* to appear in Bull. Math. Sci. Bull. Sci. Math. 137(2013), 509–522.
9. J. M. Bismut, *Large Deviations and the Malliavin Calculus,* Progress in Mathematics 45, Birkhäuser, Boston, MA, 1984.
10. G. Da Prato, M. Röckner, F.-Y. Wang, *Singular stochastic equations on Hilbert spaces: Harnack inequalities for their transition semigroups,* J. Funct. Anal. 257(2009), 992–1017.
11. G. Da Prato, J. Zabczyk, *Stochastic Equations In Infinite Dimensions,* Cambridge University Press, Cambridge, 1992.
12. G. Da Prato, J. Zabczyk, *Ergodicity for Infinite-Dimensional Systems,* Cambridge University Press, Cambridge, 1996.
13. B. Driver, *Integration by parts for heat kernel measures revisited,* J. Math. Pures Appl. 76(1997), 703–737.
14. K. D. Elworthy, X.-M. Li, *Formulae for the derivatives of heat semigroups,* J. Funct. Anal. 125(1994), 252–286.
15. A. Es-Sarhir, M.-K. v. Renesse, M. Scheutzow, *Harnack inequality for functional SDEs with bounded memory,* Electron. Commun. Probab. 14(2009), 560–565.
16. X.-L. Fan, *Harnack inequality and derivative formula for SDE driven by fractional Brownian motion,* Science in China—Mathematics 561(2013), 515–524.
17. X.-L. Fan, *Derivative formula, integration by parts formula and applications for SDEs driven by fractional Brownian motion,* arXiv:1206.0961.
18. S. Fang, T. Zhang, *A study of a class of stochastic differential equations with nonLipschitzian coefficients,* Probab. Theory Related Fields, 132(2005), 356–390.

F.-Y. Wang, *Harnack Inequalities for Stochastic Partial Differential Equations,* SpringerBriefs in Mathematics, DOI 10.1007/978-1-4614-7934-5,

19. A. Guillin, F.-Y. Wang, *Degenerate Fokker-Planck equations: Bismut formula, gradient estimate and Harnack inequality,* J. Differential Equations 253(2012), 20–40.
20. R. S. Hamilton, *The Harnack estimate for the Ricci flow,* J. Differential Geom. 37(1993), 225–243.
21. A. Harnack, *Die Grundlagen der Theorie des logarithmischen Potentiales und der eindeutigen Potentialfunktion in der Ebene,* V. G. Teubner, Leipzig, 1887.
22. N. Ikeda, S. Watanabe, *Stochastic Differential Equations and Diffusion Processes (Second Edition),* North-Holland, Amsterdam, 1989.
23. H. Kawabi, *The parabolic Harnack inequality for the time dependent Ginzburg-Landau type SPDE and its application*, Potential Anal. 22(2005), 61–84.
24. N.V. Krylov, B.L. Rozovskii, *Stochastic evolution equations,* Translated from Itogi Naukii Tekhniki, Seriya Sovremennye Problemy Matematiki, 14(1979), 71–146, Plenum Publishing Corp. 1981.
25. G. Q. Lan, *Pathwise uniqueness and nonexplosion of SDEs with nonLipschitzian coefficients,* Acta Math. Sinica (Chin. Ser.) 52(2009), 731–736.
26. P. Li, S.-T. Yau, *On the parabolic kernel of the Schrödinger operator,* Acta Math. 156(1986), 153–201.
27. W. Liu, *Harnack inequality and applications for stochastic evolution equations with monotone drifts,* J. Evol. Equ. 9(2009), 747–770.
28. W. Liu, *Ergodicity of transition semigroups for stochastic fast diffusion equations,* Front. Math. China 6(2011), 449–472.
29. W. Liu, M. Röckner, *SPDE in Hilbert space with locally monotone coefficients,* J. Differential Equations 259(2010), 2902–2922.
30. W. Liu, M. Röckner, *Local and global well-posedness of SPDE with generalized coercivity conditions,* J. Funct. Anal. 254(2013), 725–755.
31. W. Liu, F.-Y. Wang, *Harnack inequality and strong Feller property for stochastic fast-diffusion equations,* J. Math. Anal. Appl. 342(2008), 651–662.
32. P. Malliavin, *Stochastic Analysis,* Springer-Verlag, Berlin, 1997.
33. X. Mao, *Stochastic Differential Equations and Their Applications*, Horwood, Chichester, 1997.
34. J. Moser, *On Harnack's theorem for elliptic differential equations,* Comm. Pure Appl. Math. 14(1961), 577–591.
35. D. Nualart, *The Malliavin Calculus and Related Topics,* Springer-Verlag, New York, 1995.
36. S.-X. Ouyang, *Harnack inequalities and applications for multivalued stochastic evolution equations,* Infin. Dimens. Anal. Quantum Probab. Relat. Top. 14(2011), 261–278.
37. E. Pardoux, *Sur des equations aux dérivées partielles stochastiques monotones,* C. R. Acad. Sci. 275(1972), A101–A103.
38. E. Pardoux, *Equations aux dérivées partielles stochastiques non lineaires monotones: Étude de solutions fortes de type Ito,* Thèse Doct. Sci. Math. Univ. Paris Sud. 1975.
39. C. Prévôt, M. Röckner, *A Concise Course on Stochastic Partial Differential Equations,* Lecture Notes in Mathematics 1905, Springer, Berlin, 2007.
40. J. Ren, M. Röckner, F.-Y. Wang, *Stochastic generalized porous media and fast diffusion equations,* J. Differential Equations 238(2007), 118–152.
41. M. Röckner, F.-Y. Wang, *Harnack and functional inequalities for generalized Mehler semigroups,* J. Funct. Anal. 203(2003), 237–261.
42. M. Röckner, F.-Y. Wang, *Non-monotone stochastic generalized porous media equations,* J. Differential Equations 245(2008), 3898–3935.
43. M. Röckner, F.-Y. Wang, *Log-Harnack inequality for stochastic differential equations in Hilbert spaces and its consequences,* Infin. Dimens. Anal. Quantum Probab. Relat. Top. 13(2010), 27–37.
44. M.-K. von Renesse, M. Scheutzow, *Existence and uniqueness of solutions of stochastic functional differential equations,* Random Oper. Stoch. Equ. 18(2010), 267–284.
45. T. Seidman, *How violent are fast controls?* Math. of Control Signals Systems, 1(1988), 89–95.
46. J. Serrin, *On the Harnack inequality for linear elliptic equations,* J. Analyse Math. 4(1955/56), 292–308.

47. J. Shao, F.-Y. Wang. C. Yuan, *Harnack inequalities for stochastic (functional) differential equations with nonLipschitzian coefficients,* Elec. J. Probab. 17(2012), 1–18.
48. T. Taniguchi, *The existence and asymptotic behaviour of solutions to nonLipschitz stochastic functional evolution equations driven by Poisson jumps,* Stochastics 82(2010), 339–363.
49. A. Thalmaier, *On the differentiation of heat semigroups and Poisson integrals,* Stochastic Stochastic Rep. 61(1997), 297–321.
50. F.-Y. Wang, *Logarithmic Sobolev inequalities on noncompact Riemannian manifolds,* Probab. Theory Related Fields 109(1997), 417–424.
51. F.-Y. Wang, *Equivalence of dimension-free Harnack inequality and curvature condition,* Integral Equations Operator Theory 48(2004), 547–552.
52. F.-Y. Wang, *Functional Inequalities, Markov Semigroups and Spectral Theory,* Science Press, Beijing, 2005.
53. F.-Y. Wang, *Dimension-free Harnack inequality and its applications,* Front. Math. China 1(2006), 53–72.
54. F.-Y. Wang, *Harnack inequality and applications for stochastic generalized porous media equations,* Ann. Probab. 35(2007), 1333–1350.
55. F.-Y. Wang, *On stochastic generalized porous media and fast-diffusion equations,* J. Shandong Univ. Nat. Sci. 44(2009), 1–13.
56. F.-Y. Wang, *Harnack inequalities on manifolds with boundary and applications,* J. Math. Pures Appl. 94(2010), 304–321.
57. F.-Y. Wang, *Harnack inequality for SDE with multiplicative noise and extension to Neumann semigroup on nonconvex manifolds,* Ann. Probab. 39(2011), 1449–1467.
58. F.-Y. Wang, *Coupling and applications,* in "Stochastic Analysis and Applications to Finance" (edited by Tusheng Zhang and Xunyu Zhou), pp. 411–424, World Scientific, 2012.
59. F.-Y. Wang, *Analysis for Diffusion Processes on Riemannian Manifolds,* World Scientific, Singapore, 2013.
60. F.-Y. Wang, *Derivative formula and gradient estimates for Gruschin type semigroups,* to appear in J. Theo. Probab.
61. F.-Y. Wang, *Integration by parts formula and shift Harnack inequality for stochastic equations,* arXiv:1203.4023.
62. F.-Y. Wang, *Derivative formula and Harnack inequality for jump processes,* arXiv:1104.5531.
63. F.-Y. Wang, J. Wang, *Harnack inequalities for stochastic equations driven by Lévy noise,* arXiv:1212.0405.
64. F.-Y. Wang, J.-L. Wu, L. Xu, *Log-Harnack inequality for stochastic Burgers equations and applications,* J. Math. Anal. Appl. 384(2011), 151–159.
65. F.-Y. Wang, L. Xu, *Derivative formula and applications for hyperdissipative stochastic Navier-Stokes/Burgers equations,* to appear in Infin. Dimens. Anal. Quantum Probab. Relat. Top.
66. F.-Y. Wang, L. Xu, *Log-Harnack inequality for Gruschin type semigroups,* to appear in Rev. Matem. Iberoamericana.
67. F.-Y. Wang, L. Xu, X. Zhang, *Gradient estimates for SDEs driven by multiplicative Lévy noise,* arXiv:1301.4528.
68. F.-Y. Wang, C. Yuan, *Harnack inequalities for functional SDEs with multiplicative noise and applications,* Stochastic Process. Appl. 121(2011), 2692–2710.
69. F.-Y. Wang, T. Zhang, *Log-Harnack inequality for mild solutions of SPDEs with strongly multiplicative noise,* arXiv:1210.6416.
70. F.-Y. Wang, X. Zhang, *Derivative formula and applications for degenerate diffusion semigroups,* to appear in J. Math. Pures Appl. 99(2013), 726–740.
71. T. Yamada, S. Watanabe, *On the uniqueness of solutions of stochastic differential equations,* J. Math. Kyoto Univ. 11(1971), 155–167.
72. S.-Q. Zhang, *Harnack inequality for semilinear SPDEs with multiplicative noise,* Statist. Probab. Lett. 83(2013), 1184–1192.
73. S.-Q. Zhang, *Shift Harnack inequality and integration by part formula for semilinear SPDE,* arXiv:1208.2425.

74. T. Zhang, *White noise driven SPDEs with reflection: strong Feller properties and Harnack inequalities,* Potential Anal. 33(2010), 137–151.
75. X.-C. Zhang, *Stochastic flows and Bismut formulas for stochastic Hamiltonian systems,* Stochastic Process. Appl. 120(2010), 1929–1949.
76. X.-C. Zhang, *Derivative formulas and gradient estimates for SDEs driven by α-stable processes,* Stochastic Process. Appl. 123(2013), 1213–1228.

Index

coupling by change of measure, 1
cylindrical Brownian motion, 8

equation
  p-Laplacian, 30
  nonlinear monotone, 27
  stochastic fast-diffusion, 28
  stochastic heat, 28
  stochastic porous medium, 28

formula
  Bismut, 3
  integration by parts, 6

geodesic
  minimal, 12

inequality
  entropy–cost, 23
  gradient-L^2, 17
  gradient–entropy, 12
  gradient–gradient, 16
  Harnack, 2
  Harnack with power, 2
  log-Harnack, 2
  shift Harnack, 6

kernel, density, 20

measure
  invariant, 20
  quasi-invariant, 20

operator
  Feller, 20
  Markov, 2
  strong Feller, 20

solution
  mild, 51
  segment, functional, 80
  strong, variational, 28
  weak, 80
space
  geodesic, 12
  length, 12

F.-Y. Wang, *Harnack Inequalities for Stochastic Partial Differential Equations*,
SpringerBriefs in Mathematics, DOI 10.1007/978-1-4614-7934-5,